2ND EUROPEAN SYMPOSIUM ON ENGINEERING CERAMICS

Proceedings of the 2nd European Symposium on Engineering Ceramics, held at the Park Lane Hotel, London, 23–24 November 1987. Organised by IBC Technical Services Ltd, in association with the International Journal of High Technology Ceramics.

2ND EUROPEAN SYMPOSIUM ON ENGINEERING CERAMICS

Edited by

F. L. RILEY

Division of Ceramics, School of Materials, University of Leeds, UK

ELSEVIER APPLIED SCIENCE
LONDON and NEW YORK

ELSEVIER SCIENCE PUBLISHERS LTD
Crown House, Linton Road, Barking, Essex IG11 8JU, England

Sole Distributor in the USA and Canada
ELSEVIER SCIENCE PUBLISHING CO., INC.
655 Avenue of the Americas, New York, NY 10010, USA

WITH 32 TABLES AND 148 ILLUSTRATIONS

British Library Cataloguing in Publication Data

European Symposium on Engineering Ceramics
(2nd: 1987: London, England)
2nd European Symposium on Engineering
Ceramics
1. Materials: Ceramics. Engineering aspects
I. Title   II. Riley, F. L.
620.1'4

ISBN 1-85166-295-2

Library of Congress Cataloging-in-Publication Data

European Symposium on Engineering Ceramics (2nd: 1987: London, England
2nd European Symposium on Engineering Ceramics/editor, F. L. Riley.
    p. cm.
Proceedings of the 2nd European Symposium on Engineering Ceramics,
held at Park Lane Hotel, London, 23-24 November 1987.
Bibliography: p.
Includes index.
ISBN 1-85166-295-2
1. Ceramics—Congresses.   2. Ceramic materials—Congresses.
I. Riley, F. L.   II. Title.
TP786.E97 1987
666—dc19   88-28689

Special regulations for readers in the USA

Phototypesetting by Tech-Set, Gateshead, Tyne & Wear.
Printed in Great Britain at the University Press, Cambridge.

Preface

This volume contains the proceedings of the 2nd European Symposium on Engineering Ceramics held in London, 23–24 November 1987. The meeting was attended by almost 200 scientists and engineers, primarily drawn from industry, and the Sessions were chaired by Mr Eric Briscoe, past President of the Institute of Ceramics. Very effective symposium organisation was provided by IBC Technical Services Ltd.

The engineering ceramics are a class of materials which has over some 50 years found well-established applications based on the materials' chemical stability and wear resistance. The last 20 years have seen intensified efforts to extend applications for these materials into areas traditionally occupied by metals, but in which the typical metallic weaknesses of wear, and of high temperature creep and oxidation, are now creating significant problems. These efforts have, however, in many cases been undermined on the one hand by the inherent ceramic weaknesses of brittleness and flaw sensitivity, and on the other by an inadequate understanding, and control, of the basic ceramic fabrication processes required for the low-cost mass production of relatively complex components. The positive results of the efforts of the last 20 years have been the development of a large new group of ceramic materials believed to possess intrinsic mechanical property advantages, of which the transformation toughened zirconias, and the ceramic matrix composites are good examples, together with improved powder production methods and powder shaping processes. This activity has taken place against the background of tightly controlled cost structures for established components produced from traditional metallic materials, by comparison with which the ceramic equivalent must be seen to give value for money, or an unmatchable performance.

The vision of super-hard, wear-resistant materials, possessing high strength, and creep and oxidation resistance at extraordinarily high temperatures — to 2000 °C and above — is a tantalising one, which has stimulated much advanced materials research into very difficult areas. Attainment of complete success always seems a little further away than had been thought, though many small but significant advances are being made which on balance offer encouragement rather than the contrary. Part of the problem of the development of acceptance of ceramics by engineers has been the need to improve understanding of the materials' behaviour. It is also necessary to make clear the limitations to the technique of simply replacing an existing metal component by one made of, or coated by, a ceramic. An intelligent, carefully considered, approach to the use of this increasing range of materials options is called for.

Development activity in the ceramics field can only be fruitful if there is good collaboration between producer and potential user. Most major materials users realise this and are anxious to keep abreast of developments. For this reason a symposium of the broader review, less highly specialised, type has a place in today's crowded international conference programme. The Engineering Ceramics Symposium was planned on the assumption that current and potential users of the engineering ceramics did indeed wish to hear presentations at a high level, given by speakers representing companies and organisations in the forefront of the various development activities. Care was chosen to include in the programme speakers able to present authoritative overviews of recent work, and of probable trends in the immediate future. These speakers were also likely (and indeed were so urged) to provide realistic pictures, taking a broad perspective view of each topic. The extent to which the speakers succeeded in this far from easy task can be judged from the informative and critical assessments provided by this volume of the Symposium presentations. These will provide the interested, scientifically educated, but non-specialist reader, with a good view of development work over the 2-year period since the first European Symposium on Engineering Ceramics in 1985. It is hoped that simultaneously some useful assessment of likely trends in the next few years can be made.

In the space of this short but intensive Symposium it was not possible to present equally all aspects of this large subject. A guiding theme was 'materials' maturity', assessed in terms such as availability, quality control, properties realised, and cost-effectiveness. Broader presentations of global developments in the subject were provided, appropriately by Dr J. G. Wurm of the Commission of the European Communities, and by Professor R. C. Bradt and Professor H. Suzuki, who covered the further west, and east, respectively. Other speakers were invited to deal with progress in groups of materials, and in processing or evaluation. More detailed treatment of specific materials, or applications, had on the whole to be avoided because of programme timetabling restrictions. An exception was made here for heat-engine applications of ceramics, which because of the exacting demands made on the materials, continue to be a major point of interest, and a stimulus for much general development work. They also serve to provide a most stringent test of maturity in any material.

F. L. RILEY

Contents

List of Contributors

P. BOCH
Ecole Nationale Supérieure de Céramique Industrielle, 47–73 Avenue Albert-Thomas, 87065 Limoges, France

R. C. BRADT
Mackay School of Mines, University of Nevada-Reno, Reno, Nevada 89557, USA

B. CALES
Céramiques Techniques Desmarquest, 2 Avenue Albert-Einstein, 78190 Trappes, France

F. CAMBIER
Centre de Recherches de l'Industrie Belge de la Céramique, 4 Avenue Gouverneur Cornez, 7000 Mons, Belgium

J.-C. GLANDUS
Ecole Nationale Supérieure de Céramique Industrielle, 47–73 Avenue Albert-Thomas, 87065 Limoges, France

K. GOEBBELS
Tiede Rissprüfanlagen GmbH, Bahnhofstrasse 94–98, D-7087 Essingen bei Aalen, FRG

J. HEINRICH
Hoechst CeramTec AG, Wilhelmstrasse 14, D-8672 Selb, FRG

xi

J. HUBER
   Hoechst CeramTec AG, Wilhelmstrasse 14, D-8672 Selb, FRG

H. KNOCH
   Elektroschmelzwerk Kempten GmbH, Postfach 1526, D-8960 Kempten, FRG

I. KVERNES
   IK Technology AS, Scheigaadgt 34F2, PO Box 3741, Gamlebyen 0135, Oslo 1, Norway

Y. LINDBLOM
   IK Technology AS, Scheigaadgt 34F2, PO Box 3741, Gamlebyen 0135, Oslo 1, Norway

E. LUGSCHEIDER
   Institute of Material Science, Technische Hochschule Aachen, 5100 Aachen, FRG

G. W. MEETHAM
   Materials Research, Rolls-Royce Ltd, PO Box 31, Derby DE2 8BJ, UK

B. G. NEWLAND
   Morgan Metroc Ltd, Bewdley Road, Stourport-on-Severn, Worcestershire DY13 8QR, UK

D. A. PARKER
   T & N Technology Ltd, Cawston House, Cawston, Rugby, Warwickshire CV22 7SA, UK

F. PLATON
   Ecole Nationale Supérieure de Céramique Industrielle, 47–73 Avenue Albert-Thomas, 87065 Limoges, France

H. SUZUKI
   Emeritus Professor, Tokyo Institute of Technology, 1-12-1 O-okoyama Meguro-ku, Tokyo 152, Japan

J. TIRLOCQ
Centre de Recherches de l'Industrie Belge de la Céramique, 4 Avenue Gouverneur Cornez, 7000 Mons, Belgium

J. G. WURM
Materials Technology Service, Commission of the European Communities (DG XII), Rue de la Loi 200, B-1049 Brussels, Belgium

1

Ceramic Research Co-operation within the European Communities

J. G. WURM

Commission of the European Communities, Brussels, Belgium

1. INTRODUCTION

The Community R and D activities are in many ways complementary to the research activities of the Member States.

Community research is conducted at several levels with differing amounts of financial aid from the research budget:

(i)   by *intramural research* at the Joint Research Centre;
(ii) by *contract research* involving financial contributions from the contractor;
(iii) through transfrontier cooperation in the form of *coordinated* research among the Member States.

The impact of Community R & D is having an increasingly important effect on European science and technology e.g. with such programmes as the following ones:

— ESPRIT : Information Technology
— RACE   : Telecommunication Technology
— BRITE : Industrial Technology
— EURAM: European Research on Advanced Materials

2. INTRAMURAL PROGRAMMES — THE JOINT RESEARCH CENTRE

2.1 The Petten Establishment

Since 1976, the European Commission has conducted a programme

1

of research into 'High Temperature Materials' at the Joint Research Centre Establishment in Petten, Netherlands. In its intramural research projects, the programme has primarily addressed the fundamental mechanisms governing the behaviour of heat-resisting alloys operating in typical industrial processes.

In this respect, the Petten Laboratory has become an established centre for high temperature materials testing with extensive facilities for studies of high temperature corrosion and surface protection phenomena and thermomechanical behaviour such as creep, fatigue and thermal fatigue, conducted *in situ* within typical process environments. Thus the combined effects of stress and corrosion (simulating actual process situations) may be accurately reproduced.

Up to 1984 the programme focussed attention on metallic systems and coatings. With the start of the 1984–87 programme period an additional project on *engineering ceramics* was initiated, in recognition of the greatly increased prospects of widespread industrial exploitation of these materials. The primary objectives of this project are to clarify the mechanisms by which advanced engineering ceramics deteriorate in dynamic, mechanical and corrosive environments and to develop guidelines for material improvement. The project is formally organized along the following lines:

(i)   Study of the high temperature corrosion mechanisms of engineering ceramics.

(ii) Determination of mechanical behaviour in high temperature corrosive media.

(iii) Establishment of the interdependence between processing parameters, microstructure, and corresponding properties.

2.2 The Karlsruhe Establishment

Although a little outside the scope of this paper it should be noted that the Karlsruhe Establishment of the Joint Research Centre has built up considerable experience and a reputation in nuclear ceramics in connection with advanced nuclear fuel research. This ceramic expertise (phase diagram determination, diffusion data and theory in relation to sintering and creep mechanisms, quantitative ceramography, etc.) is expected to provide a significant input to future Commission programmes in engineering ceramics.

3. CONTRACT RESEARCH OR SHARED COST PROGRAMMES

3.1 Highlights of the First Ceramics R and D Programme 1982–85

This first Ceramics R and D programme of the Commission was approved for the period 1982–85, and was concerned with ceramic materials development and processing. It ranged from improved refractory products to advanced technical ceramics, with emphasis on the latter, both in terms of funding level and number of projects.

Work on advanced refractories mainly involved research on the use of sialon refractory applications. Two studies concentrated on the production and application of sialon materials produced by nitriding clays. Another study was concerned with the effect of sialon additions to existing refractories either in the form of powder, mixed with refractories prior to firing, or by application of sialon coatings on conventional refractories.

In the area of advanced engineering ceramics, the projects included silicon nitride based ceramics, silicon carbide, alumina, zirconia and composites.

The work on silicon nitrides was directed towards an effort on the production on powders by nitriding of silicon and by carbo-reduction/nitriding of silica. This study included also the processing parameters of reaction bonded silicon nitride.

Another project was developing new pressureless sintered silicon nitrides with the aim to improve the high temperature strength. Analogous work was aiming for an optimization of pressureless sintered, yttria-containing sialon, with respect to oxidation resistance and high temperature mechanical properties.

Research on silicon carbide was less concerned with materials development but rather with an optimization of the injection moulding technique for the production of turbine components.

Work on partially stabilised zirconia was focused mainly on production of powders and their subsequent evaluation with respect to sintering behaviour for diesel engine applications.

As far as ceramic composites are concerned, studies on dispersion strengthened composites based on zirconia and also alumina have been studied. Another group was involved in a study on preparation of mullite based composites using a reaction-sintering process between zirconia and alumina. Processing of alumina for electrical/electronic applications was part of a joint British–German effort, particularly in the study of high purity active powders. Another project was dealing with tape casting and sintering of alumina substrates.

Development of composites ceramics based on ceramic matrices and ceramic fibres received increased attention. Ceramic based fibre reinforced composites for thermo-mechanical applications was the subject of another collaborative project.

Three projects were concerned with non-destructive testing (NDT) methods including the use of high frequency ultrasonic waves, the measurement of sonic wave velocity as a means to monitor cracking during processing, and finally the development of X-ray methods.

3.2 The Advanced Materials Programme (1987–91): EURAM: European Research on Advanced Materials

This new programme covers a period of 36 months and runs from 1987 to 1991. It is concerned with science and technology of advanced materials. The programme has been formally launched with a total budget of 60 Mecu of which 50% will be contributed by the EC. Advanced ceramics R and D represents an important part of EURAM (about 30% in terms of project titles). The programme topics involving ceramics are as follows:

— optimization of engineering ceramics for use in engines;
— study of the metal/ceramic interface;
— study of ceramic composites with fibres or whiskers as reinforcement;
— basic study of high-temperature behaviour (friction measurements).

3.3 BRITE: Basic Research in Industrial Technologies for Europe

In 1983 hundreds of companies throughout the Community were asked by the EEC Commission what areas of industrial R and D they themselves considered particularly important. Nearly 700 individual items were put forward by these companies, from which nine technological areas have emerged as priorities for action. These form the backbone of the BRITE programme.

Underlying the nine technological priority areas is one primary necessity: to get research results out of the laboratories and into the factories. EEC countries are slow to ensure the industrial application of academic research even within their own frontiers; cooperation between researchers and manufacturers, and between industries, across frontiers in Europe is of primary importance.

Consequently, two general principles govern the selection of projects within the BRITE programme. If cooperative industrial R and D is to be effective, a sufficient 'critical mass' of resources must be devoted to it. At the same time, BRITE is selective, aiming at the key areas identified by industry and proposing Community cooperation only where there is real additional value — a catalyst effect — to be had from the European dimension. The BRITE programme started in 1985 with 95 selected projects. Among these projects 7 are concerned with advanced ceramics in various areas:

— wear properties;
— laser surface treatment of ceramics;
— metal/ceramic interface systems;
— sintering of Si_3N_4,
— reinforced ceramics, etc.

4. COORDINATED RESEARCH PROGRAMMES (COST)

One of the most important forms of European coordinated research involving concerted action is 'COST', European Cooperation in the Field of Scientific and Technical Research. COST forms a framework and forum for European research cooperation which includes the countries of the European Community and other European OECD Member States. Two of the COST projects involve aspects of advanced ceramics:

COST 501 — High Temperature Materials for Conventional Systems of Energy Generation and Conversion using Fossil Fuels
COST 503 — Powder Metallurgy.

The main COST 501 effort related to high tech ceramics is devoted to R and D of ceramic thermal barrier coatings for blades, valves, combustion chambers, etc. In all, over 10 projects are concerned with this area of R and D. The COST 503 topic, which involves high tech ceramics is 'Powder Metallurgy of Hard Materials, Heavy Alloys and Ceramic Materials'. There are some 10 R and D projects involving ceramic materials — mainly concerned with cutting tools.

5. EUREKA

A new vehicle for coordinated research in Europe is EUREKA. A number of EUREKA projects are concerned with advanced ceramics such as the development of fibre-reinforced ceramics for diesel engine applications. It is expected that the Commission will participate in those EUREKA projects which relate to current Community R and D interests.

6. CONCLUSION

In conclusion, it should be emphasized that although the Commission's ceramics R and D programmes are described here in a compartmentalized fashion, i.e. intramural, extramural and coordinated actions, in practice there are many formal and informal cross-linkages between the programmes. This tendency towards greater integration of programmes, which in some cases may include joint management of the intramural and extramural projects, is embodied in the Communities' Framework Programme of Technological Research and Development.

2

Pre-standardization Studies as a Means of Helping the Development of Engineering Ceramics

PHILIPPE BOCH, JEAN-CLAUDE GLANDUS AND F. PLATON

Ecole Nationale Supérieure de Céramique Industrielle, Limoges, France

ABSTRACT

The dialogue between ceramics producers and ceramics users is difficult because of the lack of standards. VAMAS is one of the international programmes devoted to pre-standardization studies. It focuses on four topics: strength and delayed fracture, wear and friction properties, hardness, and thermal shock resistance.

1. INTRODUCTION

Numerous studies have been devoted to engineering ceramics over the last two decades, in particular in the USA, Japan and the European countries (mainly in France, Germany, Sweden and the UK). The materials initially studied were the covalent, silicon-based, non-oxide materials, viz. silicon nitride (Si_3N_4), silicon oxynitrides (the so-called sialon compounds), and silicon carbide (SiC). Subsequently, numerous efforts were devoted to toughened zirconia, e.g. partially stabilized zirconia (PSZ) or tetragonal zirconia polycrystal (TZP), and zirconia-based systems (ZrO_2-alumina or ZrO_2-mullite). At the present time, the short-fibre reinforced composites, for example SiC whiskers in an alumina matrix, and the continuous-fibre reinforced composites, for example 'Nicalon' SiC fibres in a SiC matrix, are the subject of intensive study.

Engineering ceramics have been developed in order to obtain good

7

mechanical properties at high temperatures. However, each compound shows a specific behaviour, which means that these ceramic materials are more complementary partners than fierce competitors. Silicon nitride and silicon carbide ceramics can retain a significant strength at high temperature (about 500 MPa at 1400°C), but they are not free from corrosion problems in oxidizing atmospheres, in particular in the presence of contaminants. A key difference between Si_3N_4 and SiC is that the former has a very low electrical conductivity and rather a low thermal conductivity whereas the latter is a semi-conductor and a good thermal conductor. ZrO_2-based ceramics can exhibit impressive room-temperature strength and toughness (more than 2000 MPa and 10 MPa m$^{1/2}$, respectively). However, their mechanical properties can be severely degraded after annealing in wet atmospheres at moderate temperatures (around 300°C). Zirconia has a density which is about twice that of Si_3N_4 and SiC (6 g cm^{-3} instead of less than 3·5 g cm^{-3}), a very low thermal conductivity (about 1 W m^{-1} K^{-1}), and a high thermal expansion coefficient (about 10^{-5} K^{-1}). This means that zirconia ceramics are sensitive to thermal shock, even if they can be used in their fully stabilized form as plasma-sprayed thermal barriers on metal substrates. Finally, short-fibre and continuous-fibre ceramic composites show considerable promise, in particular because of their high toughness and their low notch-sensitivity. However, they are still very expensive, difficult to produce, and sensitive to high temperature corrosion.

Heat engines were the initial targets for the use of engineering ceramics. High temperature automotive turbines (T_{max} around 1400°C) and reciprocating engines, in particular diesels with T_{max} around 1000°C, were considered for the replacement of metallic parts by ceramic ones.

For turbines, the three main objectives were:

(i)   an increase in the working temperature, and thereby an increase in thermal efficiency;
(ii) the replacement of expensive and 'strategic' cobalt-based superalloys by ceramics produced from non-strategic raw materials, hence assumed to be producible at an acceptable cost;
(iii) a weight reduction due to the low density of ceramics in comparison with metals.

For diesels, the two main objectives were:

(i)   a high degree of thermal insulation, thereby an increase in thermal efficiency;

(ii) the possibility of using low-grade oils.

The huge markets of the car industries suggested very optimistic figures, amounting to billions of US dollars, which were to increase exponentially for an indefinite time!

The First European Symposium on Engineering Ceramics[1] was one of the first meetings where the validity of such optimistic forecasts was questioned. In fact, the 1987 market for engineering ceramics is still small, and for instance Japanese statistics indicate that the 'fine ceramics' market continues to be 80% 'functional ceramics' and only 20% 'structural ceramics' (in the widest meaning of these words). At the same time, the market for engineering ceramics does not encompass what it was supposed to do, 10 or 5 years ago. Ceramic turbines have not yet succeeded on an industrial scale, and only few reciprocating engines contain ceramic parts (precombustion chamber inserts, glow plugs, and the more spectacular turbochargers).[2] Fortunately, ceramic cutting tools, ceramic wear resistance and low friction parts, and ceramic armour plating, are now produced at an increasing rate.

This rather disappointing situation exists in spite of the fact that engineering ceramics may now be considered as mature materials, exhibiting very useful and, as previously said, complementary properties. Over the two decades in which most of the ceramic laboratories have been working on those materials, engineering ceramics have noticeably increased their strength and their toughness, as well as their resistance to thermal shock, creep and corrosion. Therefore, the way to solve this puzzling problem of how to increase the industrial use of engineering ceramics is not to deny the disappointing reality but to try to understand its causes, in order to improve the situation.

2. WHAT ARE THE CAUSES FOR THE SLOW DEVELOPMENT OF ENGINEERING CERAMICS?

At least three causes have combined to slow down the development of engineering ceramics, namely:

(i)   underestimation of difficulties;

(ii) overestimation of advantages;

(iii) difficulty of dialogue between ceramists and users, in particular because of the lack of standards.

This third cause is the topic of the present paper.

2.1 The Difficulties in Optimizing Engineering-Ceramic Materials and Engineering-Ceramic Parts Have Been Underestimated

From the point of view of the materials science, improvements in the characteristics of engineering ceramics have progressed at a slower pace than had been anticipated. For instance, the relationship between microstructure and properties is reasonably well understood at the qualitative level but not at the quantitative level. In addition, a compromise between characteristics having opposite dependences on a given parameter must often be accepted. A case history is silicon nitride, where hot-pressed materials with glassy residues at the grain boundaries exhibit high mechanical strength at room temperature but high creep rate at elevated temperature, whereas porous, glass-free reaction-bonded Si_3N_4 exhibits low strength at room temperature but low creep at elevated temperature. Finally, efforts have mainly been focused on 'instantaneous' properties, for example instantaneous strength, whereas the real problems are associated with durability, as will be discussed later.

From the point of view of materials-engineering, the situation is worse. Many difficulties have arisen out of the fabrication processes, and many difficulties continue to exist at this level. The key point, indeed, is not to be able to optimize laboratory samples, with the size typically of a medical pill, and a simple shape. It is to be able to optimize real parts, which can be large and of complex shape, and must be prepared in a reproducible manner, at a price acceptable to the commercial market for mass production. The trial-and-error method which must be followed to produce industrial parts is a complex game, where one has to change the material, then the technique to process it, then the design of the part, then back to the material, and so on. As far as cost is concerned, the crude simplifications such as 'silicon nitride is made out of silicon and nitrogen, sand is silicon-rich and air is 80% nitrogen, therefore Si_3N_4 will be produced at a very low cost' have obviously failed. 'High tech' ceramic powders are presently marketed at prices which are expressed in tens of US dollars per kilogramme, whereas the usual metallic parts for the car industries are produced at about one or two US dollars per kilogramme.

2.2 The Advantages Brought by 'Ceramization' Have Been Overestimated

Another simplistic view was that 'the higher the working temperature of a thermal engine (and/or: the better the thermal insulation), the higher the thermal efficiency'. However, Carnot's principle only applies to the 'thermodynamic' efficiency of an ideal, reversible engine. The thermal efficiency of a real, irreversible engine must take into account trivial phenomena such as lubrication, self-ignition, life-time and so on! And the beautiful figures of an increase in thermal efficiency by 10–20% seem to be an illusion. It is significant that a spectacular success in engine ceramization — the ceramic turbocharger — is claimed to improve the car's acceleration (because its inertia is lower than that of its metallic counterpart), but not to increase the thermal efficiency by a noticeable amount. Silicon carbide precombustion chambers fitted in some diesel cars allow, according to Toyota, an increase in power from 95 to 105 HP, but do not sensibly reduce fuel consumption. Finally, the present drop in oil prices does not favour fuel economy.

It can be accepted that the characteristics of engineering ceramics are exciting enough to bring the assurance that these materials will be used in more and more applications. However, it is possible that engineering ceramics will not lead to the 'revolution' which has been hoped for by many ceramists. The 'all ceramic' engine could perhaps run, but it is doubtful whether it would be of real interest, at least from the economic point of view. It is more reasonable to admit that all kinds of materials — metals, polymers and ceramics — have a complementary role to play. Therefore, an optimization of mechanical parts can be obtained by a judicious fitting of the characteristics of selected materials to some specification, and not by an enthusiastic decision to favour a given type of material.

2.3 The Need for Standards

Ceramic parts will never replace metallic ones as long as mechanical engineers continue to use non-sophisticated methods of design (e.g. 'safety factors'). Indeed, brittle materials require the systematic use of the most accurate design techniques (e.g. finite element methods), which are presently employed more and more. However, these methods cannot be used without an exact knowledge of the materials properties, for instance strength and scatter of strength.

Many studies have detailed the applicability of Weibull statistics and SPT diagrams.[3] However, it must be acknowledged firstly that these tools for designing brittle parts are employed more at the laboratory level than at the industrial level, and secondly that laboratories often make the situation worse by not considering how data on strength and related properties are used in industry. This underlines *the urgent need for standards tailored to serve users, and in particular to designers of mechanical parts.* At the same time, it is advisable to avoid the choice of too complex experimental rules. Attainment of such a balance is the object of the pre-standardization programme of VAMAS, which will now be described.

2.4 VAMAS

VAMAS (Versailles Project on Advanced Materials and Standards) is 'a scheme to stimulate the introduction of advanced materials into high-technology products and engineering structures with the overall aim of encouraging international trade therein:

(i)   through international agreement on codes of practice and performance standards;
(ii) through multilateral research aimed at furnishing the enabling scientific and meteorological base necessary to achieve agreements on standards'.[4]

VAMAS was created within the framework of the Working Group on 'Technology, Growth and Employment', which was set up by the Seven Heads of State and Government and Representatives of the European Communities at the Economic Summit of Versailles in June 1982. A number of Technical Working Parties have been launched, in particular 'Ceramics', led by France (Professor P. Boch, ENSCI, Limoges). This programme focuses on mechanical and thermo-mechanical properties of engineering ceramics. It concerns four areas, namely (i) strength and delayed fracture, (ii) hardness, (iii) wear and friction properties, and (iv) thermal shock resistance. The third topic (wear and friction properties) is closely connected to the Wear Test Methods theme led by Germany (Professor H. Czichos, BAM, Berlin). The following paragraphs will describe the four topics of the VAMAS programme on ceramics.

2.4.1 Strength and delayed fracture
Mechanical strength (σ_f) continues to be the most popular parameter used to characterize the mechanical properties of a material. Therefore, ceramic parts are generally designed using strength data, which requires

that the data be effectively applicable to the cases studied. This calls for a number of comments:

The strength of a sample (σ_f) is related to its toughness (K_c) and to the equivalent size of the critical flaw (a_c) by the relation:

$$\sigma_f = (1/Y) K_c/\sqrt{a_c}$$

where Y is a numerical parameter.

K_c can be measured with good accuracy, although its determination is generally carried out by methods which use samples with a macroscopic notch, which raises the question whether some 'R-curve' effects could affect the validity of results.[5,6] On the other hand, it is very rare that the critical flaw can be detected. This means that 'a_c' cannot be measured. Only statistical laws apply here, among which Weibull's statistics is the most often used. Therefore a knowledge of Weibull's modulus (m) is required to allow us to extrapolate the data obtained on small samples for a given load state, to the data suitable for large pieces and another load state.[3] This suggests a rule:

(i) The mechanical strength of a ceramic material should not be characterized by the value of σ_f only. The σ_f scattering should also be given, for instance in the form of Weibull's modulus (m).

The critical flaws which control the brittle fracture of a ceramic part can have various origins (surface scratch, pore, inclusion, grain boundary or phase boundary, etc.). However, they depend both on the nature of the ceramic material and on the type of fabrication technique used to prepare the part, keeping in mind that 'ceramic parts are made by applying ceramic technology to ceramic materials'. Indeed, users are generally more interested in the properties of a given part, prepared using a given route, than in the potential that the material would have shown if it had been prepared using another route. A basic law in statistics is that the choice of samples should be made in such a way that they are representative of the population they have to characterize. Unfortunately it is often the contrary for the evaluation of Weibull's modulus, *which is frequently determined using specimens prepared by a technique different from that which will be chosen to produce the real industrial pieces.* Commercial pamphlets often give values for Weibull's modulus of materials obtained for small bars, carefully polished, and prepared by cold isostatic pressing of dry powders. These experiments can lead to a m value of 15–20. However, in reality the industrial process will be injection moulding and coarse grinding, which cannot generally yield values for m greater than 10–12. Thus, a first comment is that we need strict standards on the manner of selecting the samples which are used to determine the Weibull modulus (a second comment is the necessity of verifying that the material effectively obeys Weibull's statistics). This suggests a second rule:

(ii)   Strength data should be obtained using samples cut from the part which is of interest or, at least, prepared by the same technique as this part. Particular attention should be paid to orientation effects (introduced by such techniques as hot-pressing, extrusion, or injection moulding) and to size effects (organic components pyrolysis, differential sintering, etc.).

Strength data generally concern experiments performed in clean and dry atmospheres, using high stress rates, hence corresponding to a very short time of fracture (e.g. less than 1 min). This 'inert' strength is not a useful indicator for industrial uses, which frequently involve corrosive and wet atmospheres and long operation times (e.g. 2000–10 000 h in the case of thermal engines). The 'improvement' of engineering ceramics is too often considered by scientists as a game, the goal of which is the increase of inert strength (for instance, a present target seems to be to reach a strength of 3000 MPa for TZP materials). However, it should be acknowledged firstly that such figures are of no interest to designers because they are not reliable, and because they do not allow the designers to establish a true hierarchy between concurrent commercial materials, and secondly that the route which is chosen to 'improve' materials may conflict with the achievement of more useful properties (e.g. resistance to slow crack growth). An example is the case of certain zirconia ceramics which pay the price of their higher inert strength with a higher sensitivity to corrosion in wet atmospheres.

The strength values obtained from laboratory tests are very optimistic in comparison with the static-fatigue limits which are of interest for industrial applications. This leads to a bad situation, where laboratories try to beat the record for high-strength data by using test conditions which are of no applicability, whereas industrial designers are unfortunately encouraged to overestimate the ceramic capabilities. Thus, a third rule could be:

(iii)   The reference data for strength should be measured using reasonably unfavourable conditions (i.e. wet atmosphere and long duration) instead of favourable conditions (i.e. dry atmosphere and short duration).

Finally, it must be pointed out that strength is very sensitive to temperature and that slow crack growth is thermally activated. This leads to a fourth rule:

(iv) Tests must be carried out at the temperature which is of interest for the industrial uses which are anticipated.

These four rules have been taken into account in the organization of an 'Interlaboratory Round Robin on Environmental Crack-Growth Parameters', which is organized, in the framework of VAMAS, by Drs S. W. Freiman and E. R. Fuller of the Ceramics Division of the National Bureau of Standards (NBS), USA. The purpose of the Round Robin is to assess methods of determining parameters related to environmentally-enhanced crack propagation in engineering ceramics. The inter-comparison studies measure crack-growth parameters for the water-enhanced fracture of various ceramic specimens by means of constant stressing rate tests ('dynamic fatigue tests').[7-9] These studies are conducted both with specimens containing a controlled flaw, produced by Vickers indentation, and with 'natural' specimens, containing only their intrinsic flaw distribution.

The specimens are bars (3 mm × 4 mm × 48 mm). Fifty per cent of the batch are indented in their midpoint (5 kg-loaded Vickers) whereas the others are used in their as-received state. Specimens are loaded to failure in four-point flexure on loading platens with spans of 40 mm and 10 mm (Fig. 1). The support platen and the specimen are immersed in distilled water (temperature: 22–24° C). Three different loading rates are used. The fastest corresponds to the chart pen traversing the total span of the chart paper in not less than 6 s. The slowest rate is the slowest which can be achieved by the laboratory's machine. The third rate is midway, on a logarithmic scale, between the other two loading rates.

The data collected by the participating laboratories are analysed by the Ceramics Division of the NBS. First, the flexural strengths ($\sigma_f i$) are determined by the different stressing rates (v_i). Then, a least squares curve fit of the expression:

$$\log \sigma_f i = \alpha + \beta \log v_i$$

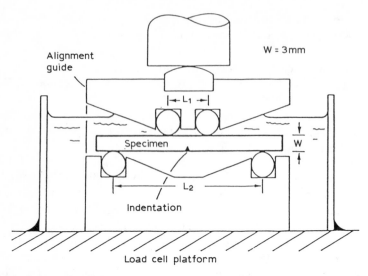

Fig. 1. Schematic of four-point flexure test on specimen immersed in water (from Ref. 9).

yields the dynamic fatigue exponent (n):

$$n = (1 - \beta)/\beta$$

The Round Robin which is currently under way involves about 20 laboratories (the EEC laboratory in Netherlands, and laboratories in France, Germany, Japan, UK and USA). The specimens are 99·7% alumina furnished by the French producer Desmarquest-Pechiney. All tests will be completed by the end of 1987, and a new Round Robin will then be carried out using silicon nitride samples, furnished by the Japanese producer NGK Spark Plug. A statistical analysis of the experimental data will be performed by the NBS, in order to define standards.

2.4.2 Hardness

Hardness testing of hard ceramics has not been standardized, and therefore VAMAS has organized a Round Robin on this topic. The Round Robin has been defined, and is coordinated, by Dr R. Morrell, Division of Materials Applications, National Physical Laboratory (NPL), Teddington, UK.

The purpose of this exercise is to evaluate the reproducibility with which different laboratories using different testers can measure hardness. It covers four different hardness tests, namely (i) Rockwell superficial HR45 (or eventually Rockwell A), (ii) Vickers macrohardness HV1 (or eventually HV2.5), (iii) Vickers microhardness HV0.2 (or eventually HV0.1), and (iv) Knoop microhardness HK0.2. Test samples are 30 mm discs of two grades of alumina. The first grade is 99·9% pure, 2 μm grain size, with a density of more than 3·95 Mg m^{-3}. It is supplied by the British producer Morgan Matroc. The second grade is a 'pink' 95% alumina, 5 μm grain size, with a density of about 3·70 Mg m^{-3}. The porosity is about 5 vol% and the second phase content is about 15 vol%. It is supplied by the British producer Lodge Ceramics.

The 99% alumina material is very hard, and its testing should reveal the limits of precision available. The 95% alumina material should reveal the difficulties that porous and multiphase ceramics pose.

The specimens supplied have a set of Vickers and Knoop indentations in them (Fig. 2). Participants of the Round Robin have to measure the size of the indentations, which have already been measured by NPL. The purpose here is to establish the reproducibility of the readings. Participants have also to make their own hardness measurements and to return the specimens to NPL to control the measurements. The purpose here is to establish machine reproducibility and the influence of the personal factor (experimenter).

An additional exercise concerns the evaluation of fracture toughness by measuring the lengths of the cracks emanating from the corners of the Vickers indentations (HV 1N).

The preliminary conclusions drawn by Dr Morrell are:[10]

(i) The majority of participants making Rockwell Superficial

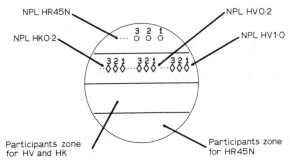

Fig. 2. Samples for the hardness tests (from Ref. 10).

HR45 indentations produced results which were very similar to those recorded by NPL. Only two results out of nine so far deviate significantly.

(ii) For HV1 Vickers hardness the range of mean results between participants was much larger than that between test discs as recorded by NPL. The differences between the NPL measurements and the participants' measurements of the same indentation was small for the fine-grained dense high-purity alumina, but rather greater for the medium-grained 95% alumina, which suffers more damage during indentation. The correlation coefficient for the two sets of measurements was generally greater than 0·5, but in some cases was negative (i.e. anticorrelated, indicating that most of the differences are random).

(iii) For HV0.2 Vickers microhardness the range of mean results from the participants was much larger than that from the same discs recorded by one operator at NPL. Differences of $> \pm 10\%$ were common, and similar results were obtained for participants reading NPL-made indentations. The correlation coefficients indicated a high degree of randomness of the data. It can be concluded that visual judgement of the measurement varies widely, and that the test is unsuited to any form of standardization.

(iv) For HK0.2 Knoop microhardness the same conclusions arise as for HV0.2.

(v) There is no correlation between eyesight and mean results.

All these comments demonstrate that further studies are required before hardness tests for engineering ceramics can be standardized. Dr Morrell is now planning another Round Robin on silicon nitride samples (supplied by the Japanese producer NGK Spark Plug).

2.4.3 Wear and friction properties

The test conditions have been defined by Professor H. Czichos (BAM, Berlin FRG: 'VAMAS wear test methods')[11] and the Round Robin has been organized by Professor H. Czichos and Dr A. W. Ruff (NBS, Gaithersburg USA). A total of 31 institutions participated (3 from Canada, 4 from FRG, 5 from France, 4 from the UK, 3 from Italy, 3 from Japan, and 9 from the USA).

Two materials were chosen for study: a 95% pure alumina, supplied by the French producer CICE, and AISI 52100 steel. The test system (Fig. 3) consists of a stationary ball (10 mm in diameter) held against a horizontally rotating disc (40 mm in the outer diameter and 32 mm in

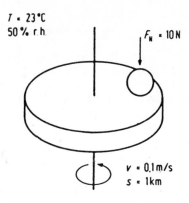

Material pairings			Measurements
disc \ ball	steel	alumina	• Friction force
steel	kit 1	kit 2	• Wear (system, ball, disc)
alumina	kit 3	kit 4	• Wear surfaces (SEM, profilometry)

Fig. 3. Sliding wear test system with specified parameters noted (from Ref. 11).

the track diameter). The test atmosphere was air, $50 \pm 10\%$ relative humidity. The test temperature was $23 \pm 1°C$. No lubricant was used. The operating variables were continuous unidirectional sliding at a velocity of 0.1 m s^{-1}, a normal load of 10 N, a sliding distance of 1 km, and a number of replicate tests of 3–5.

The specimens were used by the participants in the as-received conditions. Surfaces were cleaned immediately prior to each test by washing in freon, drying in warm air, rinsing with hexane and drying at $110°C$ for 30 min.

Measurements have concerned friction and wear properties:

(i)   Friction measurements to provide a graph showing friction force fluctuations.

(ii) Wear measurements included wear loss, wear depth, wear scar diameter on ball, and profilograms of both disc and ball surfaces. The wear surface and debris were examined by optical and scanning electron microscopy.

Four types of data were analysed by Czichos and co-workers using the reports of the participating laboratories:

TABLE 1
Mean values and standard deviations of friction and wear data (from Ref. 11)

	Steel ball steel disc	Ceramic ball steel disc	Steel ball ceramic disc	Ceramic ball ceramic disc
Coefficient of friction[a]	0·60 ± 0·11	0·76 ± 0·14	0·60 ± 0·12	0·41 ± 0·08
Number of data	109	75	64	76
Number of labs	26	26	23	26
Wear rate of system (m/km)[b]	70 ± 20	Very small	81 ± 29	Very small
Number of data	47		29	
Number of labs	11		11	
Ball wear scar diameter (mm)	2·11 ± 0·27	2·08 ± 0·35		0·3 ± 0·05
Number of data	102	[c]	60	56
Number of labs	23		21	19
Disc wear track width (mm)		0·64 ± 0·13		Not
Number of data	[d]	54	[d]	measured
Number of labs		19		

Notes:
[a]At 1000 m sliding distance.
[b]Determined from the wear curve (steady state range between 300 and 1000 m sliding distance).
[c]Material transfer from disc to ball.
[d]Material transfer from ball to disc.

Materials: AISI 52100 steel, α-Al$_2$O$_3$ ceramic
Test conditions: F_N = 10 N, v = 0·1 m/s. T = 23°C, RH 12·78%.

(i) The friction coefficient after 1 km sliding distance.
(ii) The linear wear rate defined as the displacement of the test specimen perpendicular to the sliding interface between 300 and 1000 m sliding distance divided by the sliding distance difference.
(iii) The ball wear scar diameter.
(iv) The disc wear track width.

Table 1 gives the coefficient of friction values obtained by the participating laboratories. The figures on reproducibility are:

(i) friction: ±9 to ±13% within laboratories and ±18 to ±20% between laboratories;

(ii) system wear: ±14% within laboratories and ±29 to ±38% between laboratories;
(iii) specimen wear: ±5 to 7% within laboratories and ±15 to 20% between laboratories.

The within-laboratory reproducibility ranges are said[11] to be consistent with experience at both BAM and NBS. The between-laboratory results are also consistent with those from ASTM inter-laboratory comparison.[12]

As far as the wear surfaces are concerned, Czichos's comments are:

(i) ceramic ball/steel disc: abrasion, grooving of the disc, materials transfer to the ball and tribo-oxidation of the steel;
(ii) steel ball/ceramic disc: severe abrasion of the ball, material transfer to the disc and tribo-oxidation of the steel;
(iii) ceramic ball/ceramic disc: slight abrasion of the ball and slight smoothing of the disc surface.

Czichos's conclusions on this first Round Robin exercise on wear and friction properties of alumina/steel and alumina/alumina couples are, that good within- and between-laboratory reproducibility of data can be obtained, provided the experimental conditions (and in particular the cleaning conditions) are well defined. A second VAMAS wear Round Robin is planned, using silicon nitride (supplied by Japan).

This VAMAS Round Robin exercise has focused on reproducibility, and has therefore not taken into account variations in the test parameters (e.g. temperature and atmosphere). Furthermore, most of the participating laboratories did not possess tribological machines allowing them to vary test parameters. Thus, a wear and friction cooperative programme between four laboratories[13] has been launched within the framework of the EURAM programme[14] to build a high-temperature tribometer adapted to engineering ceramics, and to study the influence of environmental parameters.[15] The tribometer (Fig. 4) has been designed to allow the variation of several parameters, in particular those related to the conditions of friction (configurations of the tribological couple) and to the conditions of environment (temperature and atmosphere). Two tribological configurations can be used:

(i) ball-on-disc configuration, which minimizes the trapping of debris;

Philippe Boch, Jean-Claude Glandus, F. Platon

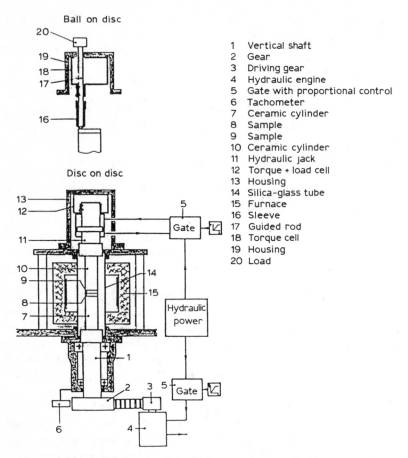

Fig. 4. Schematic of the high temperature tribometer.

(ii) a disc-on-disc configuration, which maximizes the trapping of debris.

The friction velocity can vary from 0·01 m s⁻¹ to more than 1·2 m s⁻¹ and the load can vary from 5 N to 10 N for the ball-on-disc configuration and from 200 N to 10 000 N in the disc-on-disc configuration. The temperature extends from room temperature to more than 1000° C. Tribological experiments are performed under controlled atmosphere. Different gases can be chosen, in particular air, argon and nitrogen. The relative humidity can be controlled within the 5–95% range.

The torque friction (C) is continuously recorded, in order to determine the coefficient of friction in the experiment (f) and the sliding velocity (v). The rate of wear can be determined from the loss in mass of the ball and of the disc. However, the mass losses are sometimes too small (from 10^{-4} down to 10^{-7} g) to be significant. This, in particular, is the case for the ball-on-disc configuration. Therefore, the rate of wear (k = the worn volume divided by the load and the sliding distance) is here deduced from measurement of the diameter (D) of the flat area which develops on the worn ball (observed by scanning electron microscopy). The profiles of the wear track on the disc and of the grooves on the flat area on the ball are recorded using a profilometer. Finally, the morphology of the layer debris is observed by scanning electron microscopy.

The tribological tests have concerned several materials, namely ball-bearing steel, alumina, silicon carbide, silicon nitride and partially stabilized zirconia. However, the present paper only considers the example of the influence of humidity on the wear and friction properties of alumina and silicon carbide materials, for the ball-on-disc configuration. Both ball and disc were cleaned for 5 min in an ultrasonic unit using an ethanol bath. Then they were rinsed with hexane and dried at 110°C for 30 min.

Humidity plays a significant role in the mechanical properties of ceramic materials, in particular by noticeably increasing the rate of the subcritical crack growth. Therefore, tests were conducted to examine the influence of the moisture content in air on the friction and wear behaviour of a contact consisting of an alumina ball sliding on an alumina disc. Tribological tests were carried out at room temperature under a load of 10·3 N, at a sliding velocity of 0·1 m s^{-1}, for a total sliding distance of 1 km, which corresponds to the experimental conditions of the VAMAS program. The relative humidity (RH) ranged from 5 to 95%.

Profilometer tests were carried out on the disc and on the ball, after a sliding distance of 1 km. For the disc, there was a smoothening of the surface rugosity, but not a noticeable flat worn zone. Therefore, the wear rate was evaluated from the diameter (D) of this worn zone. Both the friction coefficient and the wear rate begin to increase when humidity increases: f increases from 0·2 to 0·4 and D increases from 0·2 mm to 0·39 mm when the RH increases from 5 to 70%; for RH values beyond 70%, there is a plateau or even a decrease in f, whose value is only 0·3 when the RH = 95%. Two complementary reasons can explain why humidity increases the friction coefficient and the wear rate. First, humidity accelerates crack growth, which can lead to an increase in the production of the debris. Secondly, humidity can modify the rheological behaviour of debris, thereby modifying the protective (non-protective) influence of the layer of debris. Nevertheless, the sensitivity of f and D to RH demonstrates that standardization of tribological tests must take into account the relative humidity in the tribometer atmosphere.

2.4.4 Thermal shock resistance

Two main theories underly studies of the thermal shock resistance of brittle materials: the thermoelastic theory and the energetic theory.[16] The former assumes the thermal damage is controlled by the initiation of new cracks whereas the latter assumes that it is controlled by the propagation of pre-existing cracks. The two theories are complementary. They lead to definitions of various 'thermal shock parameters' labelled R, R', R'', R''', R'''', R_{st}, R'_{st}, etc. These parameters allow us to rank different materials for different thermal shock conditions. For instance, R deals with the critical temperature difference (dT_c) in the case of fast shock, whereas R' (where $R' = kR$, k being the thermal conductivity of the material) deals with dT_c in the case of mild shock.

Figure 5 is a flow chart which illustrates the questions to be answered to determine the thermal shock resistance of a given part, made out of a given material, and submitted to given thermal conditions.

The chart shows that the whole problem can be broken into three sequential problems, namely:

(i)   What is the temperature map inside the thermally shocked part?
(ii) What is the thermal stress map inside the part?
(iii) What is the extent of the damage inside the part?

This asks the preliminary question whether there are any useful thermal shock tests — and then whether some of them should be standardized — or whether it would not be better to standardize only primary, non-ambiguous parameters, e.g. strength (σ_f), Young's modulus (E), Poisson's ratio (v), thermal expansion (α), thermal conductivity (k), etc. Then, it would be possible to derive thermal shock parameters (for instance R, which is equal to $\sigma_f(1 - v)/\alpha E$, and R' which is equal to kR), and after that to use computerized methods to evaluate the thermal shock resistance of a given part in given experimental conditions. The best answer to this question is maybe yes, but ceramists and users are too accustomed to consider that the critical temperature difference (dT_c) is a significant parameter to accept the absence of dT_c in materials data

HEATING CONDITIONS

GEOMETRY OF THE PART

THERMAL PROPERTIES OF THE MATERIAL

1°-------> TEMPERATURE MAP INSIDE THE PART?

GEOMETRY OF THE PART

BOUNDARY CONDITIONS

ELASTIC PROPERTIES OF THE MATERIAL

THERMAL EXPANSION OF THE MATERIAL

2°------> THERMAL STRESS MAP INSIDE THE PART?

CRITERION OF DAMAGE

THERMOELASTIC and/or ENERGETIC
THEORY   THEORY
(Initiation of   (Propagation
new cracks)   of pre-existing cracks)

3°----------------> DAMAGE OR NO DAMAGE

Fig. 5. Flow chart of the problem of the thermal shock resistance of a part.

sheets! Therefore, the determination of dT_c must be standardized.

Industrial parts are generally not designed to resist to a unique and fast thermal shock but rather to endure the repetition of numerous mild thermal shocks. Therefore, thermal fatigue and slow crack growth are involved.

(i) The classical approach is that of Kamiya and Kamigaito[16] who have only considered athermal slow crack growth:

$$v = AK^n$$

They apply to thermal fatigue the formalism which is currently applied to mechanical cyclic fatigue. Thus, the lifetime of a part is expressed as a function of the number (N_i) of cycles of a given thermal amplitude (dT_i) that the part will have to resist. This leads to an expression which is similar to that for cyclic fatigue:

$$N_i \, (dT_i)^n = \text{constant}$$

and therefore:

$$N_1/N_2 = (dT_1/dT_2)^{-n}$$

for the cyclic lifetimes N_1 and N_2 under the shocks of respective severity dT_1 and dT_2.

(ii) A more complete approach has been proposed by Singh *et al.*,[17] who have considered thermally activated slow crack growth:

$$v = AK^n \exp(-Q/RT)$$

where Q is the activation energy and R the gas constant.

The quenching of a sample from a higher temperature (T_{maxi}) down to a lower one (T_{mini}) leads to three different cases:

(i) T_{maxi} keeps a constant value, and dT varies by varying T_{mini}

The expression given by Kamiya and Kamigaito is still valid:

$$N_1/N_2 = (dT_1/dT_2)^{-n} \tag{1}$$

hence, the lifetime is a function of the *n* parameter only.

(ii) dT keeps a constant value, and T_{maxi} and T_{mini} vary by the same quantity

The expression of the lifetime is:

$$N_1/N_2 = \exp(Q \, B/R) \tag{2}$$

with $B = 1/T_{maxi} - 1/T_{mini}$, hence, the lifetime is a function of the activation energy (Q) only.

(iii) T_{mini} keeps a constant value, and dT varies by varying T_{maxi}

The expression of the lifetime is:

$$N_1/N_2 = (dT_1/dT_2)^{-n} \exp(Q \, B/R) \tag{3}$$

hence, the lifetime is function of both *n* and *Q*.

These expressions are only valid for rather fast shocks. In the case of mild shocks, some corrections have to be made. For instance, n must be replaced by $1 \cdot 25n$.

The usual practice for thermal shock tests is to heat a sample in a furnace up to T_{maxi} and then to quench it down to a bath at T_{mini}. The operation is repeated for N cycles until irreversible damage develops inside the sample. Then, similar experiments are carried out again using other samples and other thermal cycle conditions, in order to draw the $N_i = f(dT_i)$ relationship.

Generally, T_{mini} (i.e. the temperature of the quenching bath) keeps a constant value, and the dT variations are obtained by varying T_{maxi} (i.e. the temperature of the furnace). This corresponds to the third case and therefore eqn. (3) should be used. However, it is very common to consider eqn. (1), which is incorrect because the thermal activation of slow crack growth is not taken into account.

We have built a fully automated thermal shock bench, designed to allow us to carry out experiments in the framework of Hasselman's theory (block-diagram shown in Figs. 6(a) and 6(b)). The bench consists of an electric furnace and a thermostatic bath, both being regulated in such a way that the two conditions $T_{\text{furnace}} < T_{\text{bath}}$ or $T_{\text{furnace}} > T_{\text{bath}}$ can be chosen. Therefore, increasing thermal shocks, as well as decreasing thermal shocks, can be obtained. The ceramic sample (a disc 30 mm in diameter and 3 mm thickness or a rod 60 mm in length and 6 mm in

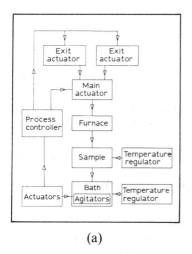

(a)

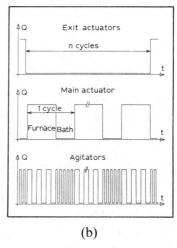

(b)

Fig. 6. Block diagram of the thermal shock bench. (a) Block diagram of the experimental device; (b) output levels of the process controller.

diameter) is gripped in a low thermal capacity metallic container, which is moved by a pneumatic actuator. Quenching of the sample from the furnace to the bath is carried out using very reproducible conditions: the velocity of the sample is controlled, and the cooling liquid is agitated in a regular manner. Various cycles can be designed, to submit the sample to successive thermal shocks, at temperatures varying from room temperature to 1000° C for the furnace and from room temperature to 400° C for the quenching bath. The heating time and the cooling time can vary from 1 to 999 s, and the number of thermal cycles can vary from 1 to 999. A preliminary study is presently under way to determine to what extent such a bench could be proposed as a reference to standardize thermal shock tests.

The assessment of damage is done by visual inspection (dye penetrant) or by breaking shocked samples and estimating the extent of the thermal cracking from the loss in strength. It could also be done using a non-destructive acoustic method. The best methods are those which allow an in situ evaluation of the extent of cracking, for instance ultrasonic methods,[18] or acoustic emission.[19] Finally, it could be instructive to use indented samples.[20, 21]

3. CONCLUSIONS

The main conclusions are:

(i) The definition of standardized tests to determine thermo-mechanical properties of engineering ceramics is considered to favour the dialogue between ceramists and users, and therefore to favour the development of these ceramics.

(ii) Prestandardization studies are necessary to define which physical parameters are of prime interest, and how they can be accurately determined.

(iii) Tests must be defined to gather quantitative information on the long term behaviour of engineering ceramic parts.

(iv) It is wrong to choose tests which give too optimistic values. It is better to choose tests which give too pessimistic values.

(v) The 'Ceramics' program of VAMAS is involved in four topics:

- Environmental crack growth parameters (led by Dr S. W. Freiman, NBS, USA).

- Hardness (led by Dr R. Morrell, NPL, UK).
- Wear and friction properties (in connection with the Wear program, led by Professor H. Czichos, BAM, FRG).
- Thermal shock resistance (led by Professor J. C. Glandus, ENSCI, France).

REFERENCES

1. *Proceedings of the First European Symposium on Engineering Ceramics*, Oyez Scientific and Technical Services, London, 1985.
2. *Proceedings of the First International Symposium on Ceramic Components for Engines*, Somiya, S., Kanai, E. and Ando, K. (eds), Tokyo, KTK Scientific Publications and London, Elsevier Applied Science Publishers, 1984; *Proceedings of the Second International Symposium on Ceramic Components for Engines*, Bunk, W. and Hausner, H. (eds), Verlag Bad Honnef, 1986.
3. Creyke, W. E. C., Sainsbury, I. E. J. and Morrell, R., *Design with non-ductile materials*, London, Applied Science Publishers, 1982.
4. VAMAS bulletin (1, January 1985; 2, July 1985; 3, January 1986; 4, July 1986; 5, January 1987; 7, July 1987; 8, to be published in January 1988). The editor was initially Dr E. D. Hondros, NPL, UK, but is now Dr B. Steiner, NBS, USA.
5. Mussler, B., Swain, M. and Claussen, N. Dependence of fracture toughness of alumina on grain size and test technique, *J. Am. Ceram. Soc.*, **65** (11) (1982) 567–72.
6. Swain, M. and Rose, L., Strength limitations of transformation-toughened zirconia alloys, *J. Am. Ceram. Soc.*, **69** (7) (1986) 511–18.
7. Gonzalez, A. C., Multhopp, H., Cook, R. F., Lawn, B. R. and Freiman, S. W. Fatigue properties of ceramics with natural and controlled flaws: a study on alumina, Special Technical Publication 844, ASTM, pp. 43–56, 1984.
8. Fett, T. and Munz, D. Time to failure in static bending tests on Al_2O_3 with natural and artificial surface flaws, *Ceramic Forum International*, **61** (9/10) (1984) 446–453.
9. Freiman, S. W. and Fuller, E. R., Ceramics testing under VAMAS: interlaboratory round robin on environmental crack growth parameters, VAMAS programme, 1987.
10. Morrell, R., VAMAS: test procedure for hardness testing of Ceramics programme, VAMAS programme, 1987.
11. Czichos, H., Becker, S. and Lexow, J. Multilaboratory tribotesting: results from the Versailles Advanced Materials and Standards Programme on wear test methods, *Wear*, **114** (1987) 109–30.
12. ASTM Committee G2 Research Reports on Standards G77 and G83, Philadelphia, USA, ASTM.
13. Study of the tribology of technical ceramics based on SiC, Si_3N_4, and ZrO_2 as a function of temperature and environment, Research action in the

EURAM programme, ENSCI Limoges, INSA Villeurbanne, Céramiques et Composites Bazet, and University of Surrey.

14. Wurm, G. J. *Ceramic programme cooperation within the European Communities*, Second European Symposium on Engineering Ceramics, London, November 23–24, 1987.

15. Kapelski, G., Platon, F. and Boch, P. Wear and friction properties of some engineering ceramics, *Science of Ceramics*, No. 14, Canterbury, September 7–9, 1987.

16. Kamiya, N. and Kamigaito, O. Prediction of thermal fatigue life of ceramics, *J. Mater. Sci.*, **14** (1979) 573–82.

17. Singh, J. P., Nihara, D. and Hasselman, D. P. H. Analysis of thermal fatigue behaviour of brittle structural materials, *J. Mater. Sci.*, **16** (1981) 2789–97.

18. Gault, C., Ultrasonic spectroscopy method for damage evaluation in ceramics submitted to thermal fatigue, *Proceedings of the Second International Symposium on Ceramic Materials and Components for Engines*, Bunk, W. and Hausner, H. (eds) Verlag Bad Honnef, 1986, pp. 869–76.

19. Glandus, J. C., Boch, P. and Jouin, C., Resistance to thermal fatigue and standards, *Science of Ceramics*, No. 13, Supplément au *Journal de Physique*, C1 (2) (47) (1986) 643–7.

20. Hasselman, D. P. H. Thermal stress resistance of engineering ceramics, *Mater. Sci. Engng*, **71** (1985) 251–64.

21. Glandus, J. C. and Boch, P., Main testing methods for thermal shocks, *Interceram*, **51** (1984) 33–7.

3

An Overview of the Movements in the Engineering Ceramics Area 1985–1987

J. TIRLOCQ AND F. CAMBIER

Centre de Recherches de l'Industrie Belge de la Céramique, Mons, Belgium

ABSTRACT

This paper consists of a review of the present position with respect to structural ceramics and their evolution in the period 1985–87. Powder and fibre production, as well as new processing techniques, are considered first. A large part of the review is devoted to new kinds of materials, and covers single compound materials of the oxide and non-oxide type, as well as particulate and whisker composites. Finally, an analysis is given of current engineering ceramics reliability.

1. INTRODUCTION

This paper aims to give an overview of progress in the structural ceramic field during the last 2 years. Of the topics examined, powder and fibre production, and new processing techniques for structural ceramics fabrication (forming, sintering, machining) are considered first.

New ceramic materials and their property improvement are then presented, considering single compound materials of the oxide and non-oxide type, and particulate and whisker composites.

The last section consists of an analysis of the reliability of engineering ceramics.

2. POWDER AND FIBRE PRODUCTION

Because engineering ceramic technology has seen rapid progress during recent years, the need for fine, and homogeneous, ceramic powders has increased. This situation is confirmed by the large number of papers devoted to powder preparation presented in major international meetings. For example the American Ceramic Society had to modify its policy, and sponsored in August 1986 a 'Conference on Ceramic Powder Science and Technology'. It also organized, with the DeutscheKeramische Gesellschaft, the 'First International Conference on Ceramic Powder Processing Science' in November 1987.[1-3]

By comparison with the traditional powder preparation routes, where usually a crushing or a milling step is necessary, the new synthesis routes aim to produce directly by controlled chemical processes very fine powders having a high degree of consistency in their main characteristics. These new routes may be classified as sol-gel and hydrothermal processes where a liquid phase is used, and as thermal processes where solid–solid or solid–gas reactions are activated by laser, or plasma energy, or by combustion enthalpy. In the sol-gel routes, the main precursors are of the oxy-chloride and alkoxide type, and freeze-drying of the gel is now preferred (for example, in the preparation of mullite, silicon carbide and tetragonal zirconia).[4-7] The laser-driven pyrolysis of organometallic compounds yields very fine and uniform powders ($<0.1\ \mu$m) and is now proposed for the production of β-silicon carbide, boron nitride and silicon nitride.[1, 8-10]

Concerning ceramic fibre production, the use of organic precursors has been reported several times, using for instance polycarbosilane to obtain silicon carbide, and polysilazane for silicon nitride production. According to Okamura and co-workers, complete nitridation of polycarbosilane below 1000° C gives rise to an amorphous fibre, which can be crystallized to α-Si$_3$N$_4$ at 1400° C, or to silicon oxynitride, depending on the annealing conditions. Tensile strengths of such fibres are high ($\sim$1.5 GPa) and show their maximum values at 1300° C for the α-Si$_3$N$_4$ fibre and at 1400° C for the oxynitride.[11]

Synthesis of an amorphous ceramic fibre (diameter 10–15 μm) of the Si-C-N type from an hydridopolysilazane polymer has also been reported, with fibre tensile strengths higher than 3.0 GPa.[12]

3. NEW PROCESSING TECHNIQUES FOR STRUCTURAL CERAMICS PRODUCTION

Stringent property requirements for engineering ceramic materials lead to new production methods at all stages of their manufacture. As far as the forming step is considered, the organometallic polymers used as precursors for powder and fibre production, as mentioned above, can also function as binders and pressing aids in the green body, and after pyrolysis form a ceramic bond in the material. For example, silicon carbide powder–polycarbosilane mixtures can be extruded, and then sintered to 97% of theoretical density. Modeling of these pre-ceramic polymer/ceramic powder systems has been studied for forming by die-pressing and extrusion.[14] Additionally several comparative studies of ceramic injection moulding formulations and parameters have been published.[15, 16]

Concerning the sintering step, most communications have been devoted to the densification of non-oxide materials. The following key points have to be noticed:

— The benefit in nitriding silicon powder compacts in a mixed nitrogen–hydrogen atmosphere, with an increase in the nitride fraction, using hydrogen contents up to 40 vol%.[17]
— The competition between densification and nitridation of silicon powder when a submicrometre raw material ($0 \cdot 1$–$0 \cdot 2 \, \mu$m diameter) synthesized by vapour-phase reaction in an aerosol is used.[18]

Gas pressure sintering of silicon nitride and sialon has given rise to many communications.[19-22] This sintering technique allows an increase in the soaking temperature, up to 1800–2000°C at pressures between 1 and 4 MPa, without decomposition of silicon nitride, and to reduce the amount of sintering additives used, in comparison with that required in normal pressureless sintering. Attractive results are reported by Tani and co-workers[19] for two stage silicon nitride sintering using a mixed additive of Al_2O_3 and rare earth oxides (Y_2O_3, La_2O_3 and CeO_2). A bulk density of $3 \cdot 27$ g/cm^3, flexural strength of 800 MPa and fracture toughness up to 9 MN m$^{-3/2}$, are obtained. This latter value is attributed to a fibre-like microstructure. Similar flexural strengths are obtained by Ueno with alumina-praseodymium oxide mixed additive.[21]

Hot isostatic pressing (hipping), the most prestigious pressure sintering technique, is reported to be useful in the improvement of reliability (more uniform density and higher Weibull modulus) of complex shapes such as gas turbine wheels,[23-25] produced from silicon nitride and silicon carbide. The hipping technique indeed allows better control of grain growth, and is able to eliminate in many cases internal laminations and voids in the powder body. Chemical impurities and hard agglomerates due to previous errors in preparation cannot of course be removed. Encapsulation is considered to be effective in preventing, by means of the impermeable container, gas escape and, as a consequence, in hindering decomposition or dissociation at sintering temperature of nitride, sialon and oxycarbide compounds.

Due to the superplastic hot deformation properties of yttrium stabilized zirconia (Y–TZP) and Al_2O_3-ZrO_2 composites, hot forging may be considered as a candidate for ceramic processing. Large strains and strain rates leading to void elimination without crack formation and phase transformation are possible, provided that the grain size is limited to 1 μm.[26, 27]

The economic production of high precision engineering components with functional surfaces, presupposes that the problems of machining materials such as silicon carbide, silicon nitride, or boron carbide, are solved. Review articles on this topic discussing, recommended bonds for diamond tools, and machining methods, have been published recently.[28, 29] Residual stresses in machined ceramic surfaces have been measured by two methods: the first consists of determining by X-ray diffraction changes in the lattice spacing of the polycrystalline material on the machined surface; the second is the measurement, by an interference method, of the bending of a thin plate, resulting from the compressive stress of the machined surface against the polished opposite face which becomes concave.[30]

3.1 Single Compound Materials of the Oxide Type

Studies on well known single oxide materials such as alumina are less common but experiments on the effects of MgO as a solid-solution additive on the sintering and microstructure development of α-Al_2O_3 have been reported. Conclusions drawn are that MgO increases the densification rate during sintering by a factor of 3 and also the grain growth rate by a factor of 2·5.[31] The influence of the degree of densification of alumina powder compacts after pressureless sintering at various temperatures on the effect of a post-hipping, has also been examined, and the difficulty of eliminating pores trapped inside the grains after sintering at too high a temperature has been demonstrated.[32]

Tetragonal zirconia polycrystal is now considered as an attractive engineering ceramic material because of its excellent mechanical properties, and a thermal expansion coefficient close to that of cast iron. Nettleship and Stevens have published (1987) an extensive review paper on this topic, and in which high toughness tetragonal zirconia polycrystals (TZP) developed in the ZrO_2–CeO_2 system are compared to the yttria stabilized crystals.[33] The remaining problem of the low-temperature ageing of TZP has been considered,[34] and new additives such as CeO_2 and TiO_2 used as a mixed additive with yttria have been tried. Sato and co-workers[35] concluded that a serious decrease in strength occurs, due to the ageing effect in humid air or hot water, with both types of mixed additive, but by alloying yttria with CeO_2, grain growth and loss of mechanical properties can be avoided.

Low thermal expansion ceramic materials of the oxide type suitable for applications involving severe thermal shock have been investigated:

— High cordierite, as a solid solution containing Fe, Mn, Ga or Ge has been synthesized,[36] and germanium-modified cordierite ceramic seems to be very promising, with regard to its near-zero thermal expansion coefficient.[37]

— Additions of BaO, SiO_2, MgF_2, as thermal stabilizers have been tried in aluminium titanate.[38] Substitution of Al by Si and 2Al by Mg and Ti was found to be effective in controlling thermal decomposition.[39]

— Aluminium titanate–mullite composites are a flexible compromise between high mechanical strength and thermal shock resistance, depending on the titanate content.[40]

— The chemically stabilized α-phase of zirconyl phosphate ($2ZrO_2.P_2O_5$ or $(ZrO)_2P_2O_7$) can constitute a dense, low-thermal-expansion, polycrystalline ceramic, with a controlled grain size.[41]

3.2 Single Compound Materials of the Non-oxide Type

Among the single compound materials of the non-oxide type silicon carbide (SiC), silicon nitride (Si_3N_4) and sialon, have been the most investigated. Concerning silicon carbide, cautions regarding the conditioning of submicrometre SiC powder before handling,[42] the negative effect of oxidized species on particulate surfaces of α-SiC,[43] thermodynamic considerations regarding effective sintering aids,[44] and the effect of sintering temperature on the properties of β-SiC,[45] have been reported. Relations between grain-boundary structure and high temperature strength with various additives, and relations between microstructure and tribological properties, have also been published.[46, 47]

In the case of silicon nitride, the marked effect of starting powder characteristics on pressureless sintered, and hipped material properties,[48-50] and the benefits of controlled crystallization of the amorphous phase for high temperature mechanical properties[51] have again been emphasized. Oxidation and creep behaviour of hot-pressed, sintered and post-sintered Si_3N_4 materials have been compared, and related to the first formation of liquid phase at high temperature.[52]

The effects of sintering additives,[53] composition (various z values),[54] and heat treatment[55] on the mechanical properties of α'- and β'-sialon materials have been reported. The best mechanical strengths at room temperature, and at 1400° C, are obtained for sialon of $z = 1$, with SiC, NbC, TiC and TiN sintering additives which subsequently act as crack propagation barriers.[51]

3.3 Particulate Composite Materials

For particulate composites with an oxide matrix, several new systems, or layer devices, have been investigated:

— Zirconia transformation toughening has been applied in the strengthening of two or three layer laminated composites, with Al_2O_3-unstabilized ZrO_2 as the outer layer and an Al_2O_3-stabilized ZrO_2, as the central layer. Die pressing or tape casting was used as the forming method.[56, 57]
— Systems such as Al_2O_3-AlON and $Al_2O_3Cr_2O_3/ZrO_2HfO_2$ show attractive properties for high temperature applications. The first, with mechanical properties similar to those of pure alumina at and 1400° C, while the second shows improvement between 300 and 1000° C.[58, 59]
— Dispersed diamond particles in an alumina matrix improve toughness after transformation to graphite by a thermal treatment.[60]
— Study of ZTA properties has shown their improved wear and thermal shock resistance when compared with classical alumina or TZP.[61] Moreover such materials present a particularly interesting improvement of the subcritical crack growth resistance.[62]

In the case of non-oxide matrices, particulate composites have been developed in several systems: SiC-TiC, SiC-TiB$_2$, Si$_3$N$_4$-SiC and Si$_3$N$_4$-diamond. Improvement of the fracture toughness by stress-induced microcracking is observed in this last system,[63] while in a SiC dispersed Si_3N_4 composite, strengthening is attributed to a decrease in the flaw size, with the fracture toughness remaining constant.[64, 65]

In TiB$_2$-dispersed in α-SiC, an increase of toughness and strength is observed, this last property being preserved up to 1200° C,[66] while for TiC dispersed in α-SiC only a slight bend strength increase is reported.[67]

Synthesis of zirconium oxycarbide and ZrO_2-dispersed ZrC_xO_y composites has been carried out by Barnier and Thevenot.[68] In these composites, the introduction of oxygen as oxycarbide, and of ZrO_2 particles, allows a decrease in the hot-pressing temperature from ~2000°C (for the pure zirconium carbide) to 1550°C.

Dispersion of ZrO_2 and TiC in a TiB_2 matrix improves the mechanical properties of these particulate composites.[69-71] Dispersion of TiB_2 in TiN and Ti (CN) leads to an improvement in frictional properties at high temperature (~800°C).[72]

3.4 Fibre and Whisker Composite Materials

In the fibre-reinforced ceramic composites field, conspicuous progress has been obtained in developing the chemical vapour infiltration technique (CVI) which reduces significantly the infiltration times for low-density fibrous structures by simultaneously utilizing a thermal gradient and a pressure gradient.[73] This technique has been applied, for instance, in the fabrication of SiC/SiC flat laminates, which show toughness values as high as 25 MPa m$^{1/2}$.[74] The effect of fibre coating (with BN for example) SiC fibres in zirconia-based matrix composites before processing, is that of a substantial improvement in strength and toughness[75]. Review papers describing fibre-reinforced ceramic composites fabricated by CVD, CVI, reaction bonding of impregnated liquid silicon, and the thermal decomposition of impregnated organosilicon polymers have been published.[76, 77]

Whisker-reinforced ceramic composites have given rise to many communications.[78] Most whiskers are of the SiC type, and the matrices are mainly Si_3N_4 and Al_2O_3, although the use of $MoSi_2$ and cordierite matrices has also been reported.[78-87] For SiC–Si_3N_4 materials, a significant increase in toughness, and a lesser increase in strength, is generally observed.[79-81] Hot-pressed SiC-Al_2O_3 composites show considerable improvement in their mechanical properties,[82, 83] better creep resistance,[84, 85] and thermal shock behaviour with a critical temperature difference of up to 900°C.[86] SiC whisker-$MoSi_2$ composites seem to be promising both in toughening the intermetallic phase (well-known as a heating element material) at temperatures below 1000°C where it is brittle, and in strengthening it at high temperature, where it becomes ductile.[87]

4. ENGINEERING CERAMICS RELIABILITY

The understanding of the low temperature mechanical properties of monolithic structural ceramics is now well-advanced, and the characteristics of the various defects and their associated fracture mechanisms are well known. The development of procedures for eliminating these defects during processing is in progress, for example in mastering particle dispersion and the elimination of hard agglomerates during the preparation of slurries, and in reducing the flaw population by using hipping.

The present difficulty is to detect the remaining defects by non-destructive evaluation. Because of the small size of the critical flaws, their direct imaging by optical, electronic or radiographic methods is generally impossible. Only larger defects such as voids, delaminations or cracks, are detectable by acoustic or X-ray microfocus equipment.[88] New techniques such as scanning laser acoustic microscopy,[89] and X-ray tomography[90] are promising for the detection of density variations inside manufactured parts.

REFERENCES

1. Colomban, Ph. Poudres sur mesure: la clé des céramiques fiables? *L'Industrie Céramique*, **815** (4) (1987) 250–1.
2. Kato, A. Study on powder preparation in Japan, *Am. Ceram. Soc. Bull.*, **66** (4) (1987) 647–50.
3. Yamada, K. Powders for high technology ceramics, *Ber. D. K. G.*, **64** (6/7) (1987) 250–5.
4. Ismail, M., Nakai, Z. and Somiya, S. Microstructure and mechanical properties of mullite prepared by the sol-gel method, *J. Am. Ceram. Soc.*, **70** (1) (1987) C-7, C-8.
5. Suzuki, H. and Saito, H. Processing of the fine mullite powder from metal-alkoxides and its sintering, *Yogyo-Kyokai-Shi*, **95** (7) (1987) 697–702.
6. Guo, C., Nakaga, Z. and Hamawo, K. Effect of drying method on mullite ceramics prepared from sol mixture, *Yogyo-Kyokai-Shi*, **94** (6) (1986) 583–9.
7. Ramme, R. and Hausner, H. Mechanical properties of ZrO_2 (2% Y_2O_3) derived from freeze-dried coprecipitated hydroxides, *Ber. D. K. G.*, **64** (1/2) (1987) 12–14.
8. Rice, G. Laser synthesis of Si/C/N Powder from 1,1,1,3,3,3 hexamethyl-disilazane, *J. Am. Ceram. Soc.*, **69** (8) (1986) C-183–C-185.
9. Suyama, Y., Marra, R. M., Haggerty, J. S. and Bowen, H. K. Synthesis of ultrafine SiC powders by laser-driven gas phase reactions, *Am. Ceram. Soc. Bull.*, **64** (10) (1985) 1356–9.
10. Rice, G. Laser-driven pyrolysis: synthesis of TiO_2 from titanium isopropoxide, *J. Am. Ceram. Soc.*, **70** (5) (1987) C-117–C-120.
11. Okamura, K., Sato, M. and Hasegawa, Y. Silicon nitride fibers and silicon oxynitride fibers obtained by the nitridation of polycarbosilane, *Ceramics International*, **13** (1987) 55–61.

12. Legrow, G. E., Lim, T. F., Lipowitz, J. and Reaoch, R. S. Ceramics from hydridopolysilazane, *Am. Ceram. Soc. Bull.,* **66** (2) (1987) 363–7.
13. Mutsudoy, B. C. Use of organometallic polymer for making ceramic parts by plastic forming techniques, *Ceramics International,* **13** (1987) 41–53.
14. Schwartz, K. B. and Rowcliffe, O. J. Modeling density contribution in preceramic polymer/ceramic powder systems, *J. Am. Ceram. Soc.,* **69** (5) (1976) C-106, C-108.
15. Edirisinghe, M. J. and Evans, J. R. G. Rheology of ceramic injection moulding formulations, *Br. Ceram. Trans. J.,* **86** (1987) 18–22.
16. Mutsudoy, B. C., Study of ceramic injection molding parameters, *Adv. Cer. Mater.,* **2** (3A) (1987) 213–18.
17. Seki, Y., Kadota, M., Kondoh, I. and Wetsuki, T. Nitridation of silicon compact in a gas mixture of nitrogen and hydrogen, *Yogyo-Kyokai-Shi,* **94** (5) (1986) 46–52.
18. Gregory, O. J. and Lee, S. B. Reaction sintering of submicrometer silicon powder, *J. Am. Ceram. Soc.,* **70** (3) (1987) C52, C55.
19. Tani, E., Umebayashi, S., Kishi, K. and Kobayashi, K. Gas-pressure sintering of Si_3N_4 with concurrent addition of Al_2O_3 and 5 wt% rare earth oxide: high fracture toughness Si_3N_4 with fiber-like structure, *Am. Ceram. Soc. Bull.,* **65** (9) (1986) 1311–15.
20. Tani, E., Nishijima, M., Ichinose, H., Kishi, K. and Ume Bayashi, S. Gas pressure sintering of Si_3N_4 with an oxide addition, *Yogyo-Kyokai-Shi,* **94** (2) (1986) 300–5.
21. Ueno, K. Gas pressure sintered nitride containing praseodymium oxide as sintering aid, *Yogyo-Kyokai-Shi,* **94** (7) (1986) 702–5.
22. Hirosaki, N. and Okada, A. Effect of N_2 gas pressure on sintering of α'- and β'-sialon, *Yogyo-Kyokai-Shi,* **95** (2) (1987) 235–9.
23. Larker, H. T. Consistency is a critical problem for ceramics, can hot isostatic pressing give an answer?, *Interceram,* **36** (1) (1987) 45–7.
24. Tegman, R. Hot isostatic pressing of silicon nitride, *Interceram,* **34** (1) (1985) 22–8.
25. Watson, G. K., Moore, T. J. and Millard, M. L. Effect of hot isostatic pressing on the properties of sintered alpha silicon carbide, *Am. Ceram. Soc. Bull,* **64** (9) (1975) 1253–6.
26. Wakai, F., Sakaguchi, S. and Kato, H. Compressive deformation properties and microstructures in the superplastic Y-TZP, *Yogyo-Kyokai-Shi,* **94** (8) (1986) 721–5.
27. Kellet, B. J. and Lange, F. F. Hot forging characteristics of fine grained ZrO_2 and Al_2O_3/ZrO_2 ceramics, *J. Am. Ceram. Soc.,* **69** (8) (1986) C-172, C-173.
28. Klocke, F. Machining of engineering functional surfaces on ceramic components, *Ber. D. K. G.,* **64** (617) (1987) 234–8.
29. De Beers Industriediamanten Machining non-oxide ceramics with diamant tools, *Interceram,* **36** (1) (1987) 50–1.
30. Johnson-Walls, D., Evans, A. G., Marshall, D. B. and James, M. R. Residual stresses in machined ceramic surfaces, *J. Am. Ceram. Soc.,* **69** (1) (1986) 44–7.
31. Berry, K. A. and Harmer, M. P. Effect of MgO solute on microstructure development in Al_2O_3, *J. Am. Ceram. Soc.,* **69** (2) (1986) 143–9.

32. Sung-Tae Kwon and Doh-Yeow Kim Effect of sintering temperature on the densification of Al_2O_3, *J. Am. Ceram. Soc.,* **70** (4) (1987) C-69, C-70.
33. Nettleship, I. and Stevens, R. Tetragonal zirconia polycrystal (TZP), a review, *Int. J. High Technol. Ceram.,* **3** (1987) 1–32.
34. Hong-Yang Lu and San-Yuan Chen Low temperature ageing of t-ZrO_2 polycrystals with 3 mol % Y_2O_3, *J. Am. Ceram. Soc.,* **70** (8) (1987) 537–41.
35 Sato, T., Ohtaki, S., Endo, T. and Chimada, M. Improvement of thermal stability of yttria-doped tetragonal zirconia polycrystals by alloying with various oxides, *Int. J. High Technol. Ceram.,* **2** (1986) 167–77.
36. Ikawa, H., Otagiri, T., Jmai, O., Urabe, K. and Udagawa, S. Thermal expansion of high cordierite and its solid solutions. *Yogyo-Kyokai-Shi,* **94** (3) (1986) 344–50.
37. Agrawal, D. K. and Stubican, V. S. Germanium modified cordierite ceramics with low thermal expansion, *J. Am. Ceram. Soc.,* **69** (12) (1986) 847–51.
38. Kajiwara, M., Sintering and properties of stabilized aluminium titanate, *Br. Ceram. Trans. J.,* **86** (3) (1987) 77–80.
39. Ishitsuka, M., Sato, T., Endo, T. and Shimada, M. Synthesis and thermal stability of aluminium titanate solid solutions, *J. Am. Ceram. Soc.,* **70**, (2) (1987) 69–71.
40. Morishima, H., Kato, Z., Wematsu, K. and Saito, K. Development of aluminium titanate–mullite composite having high thermal shock resistance, *J. Am. Ceram. Soc.,* **69** (10) (1986) C-226, C-227.
41. Yamai, I. and Oota, T. Low-thermal-expansion polycrystalline zirconyl phosphate ceramic, *J. Am. Ceram. Soc.,* **68** (5) (1985) 273–8.
42. Matje, P., Martin, K. P. and Schwetz, K. A. Contribution to the conditioning of SiC sintering powder, *Interceram,* **35** (5) (1986) 58–60.
43. Sasaki, T., Fukatsu, Y. and Iseki, T. Oxidation behaviour of submicron SiC powder and its sinterability, *Yogyo-Kyokai-Shi,* **95** (6) (1987) 646–51.
44. Negita, K. Effective sintering aids for silicon carbide ceramics: reactivities of silicon carbide with various additives, *J. Am. Ceram. Soc.,* **69** (12) (1986) C-308, C-310.
45. Williams, R. M., Juterbock, B. N., Shinozaki, S. S., Peters, C. R. and Whalen, T. J. Effects of sintering temperature on the physical and crystallographic properties of β-SiC, *Am. Ceram. Soc. Bull.,* **64** (10) (1985) 1385–9.
46. Ikuhara, Y., Kurishita, H. and Yoshinaga, H. Grain boundary and high-temperature strength of sintered SiC, *Yogyo-Kyokai-Shi,* **95** (6) (1987) 638–45.
47. Knoch, H. and Kracker, J. Sintered silicon carbide, *Ber. D. K. G.,* **64** (5) (1987) 159–63.
48. Homma, K., Okada, H. and Tatuno, T. Effects of starting powders on properties of normally sintered and HIP'ed Si_3N_4, *Yogyo-Kyokai-Shi,* **95** (3) (1987) 323–9.
49. Vandeneede, V., Leriche, A., Cambier, F., Pickup, H. and Brook, R. J. Sinterability of silicon nitride powders and characterization of sintered materials. In: *Non-oxide technical and engineering ceramics,* Hampshire, S. (ed.), London, Elsevier Applied Science Publishers, 1986, pp. 53–68.
50. Cambier, F., Leriche, A. and Vandeneede, V. Powder characterization and optimization of fabrication and processing for sintered silicon nitrides –

Part I; powder characterization. In: *Ceramic materials and components for engines*, Bunk, W. and Hausner, H. (eds), Deutsche Keramische Gesellschaft (1986), pp. 55–62.

51. Bonnell, D. A., Tseng-Ying Tien and Ruhle, M. Controlled crystallization of the amorphous phase in silicon nitride ceramics, *J. Am. Ceram. Soc.*, **70** (7) (1987) 460–5.

52. Ernstberger, W., Grathwohl, G. and Thummler, F. High temperature durability and limits of sintered and hot-pressed silicon nitride materials, *Int. J. High Technol. Ceram.*, **3** (1987) 43–61.

53. Nakamura, H., Umebayashi, S., Kishi, K., Tani, E. and Kobayashi, K. The effects of additives on bending strength of hot-pressed β-sialon with z = 1, *Yogyo-Kyokai-Shi*, **93** (4) (1985) 175–81.

54. Tani, E., Umebayashi, S., Okuzono, K., Kishi, K. and Kobayashi, K. Effect of composition on mechanical properties of β-sialon, *Yogyo-Kyokai-Shi*, **93** (7) (1985) 370–5.

55. Kishi, K., Umebayashi, S., Tani, E. and Kobayashi, K. Effect of heat treatment on strengths of β-sialon, *Yogyo-Kyokai-Shi*, **95** (6) (1987) 630–7.

56. Boch, P., Chartier, T. and Huttepain, M. Tape casting of Al_2O_3/ZrO_2 laminated composites, *J. Am. Ceram. Soc.*, **69** (8) (1986) C-191, C-192.

57. Virkar, A. V., Huang, J. L. and Cutler, R. A. Strengthening of oxide ceramics by transformation-induced stresses, *J. Am. Ceram. Soc.*, **70** (3) (1987) 164–70.

58. Bach, J. P. *et al.* Céramiques techniques à dispersoïdes alumine-zircone et alumine-oxynitrure d'aluminium γ, *L'Industrie Céramique*, **814** (3) (1987) 202.

59. Tien, T. Y., Brog, T. K. and Li, A. K. Toughened ceramics in the system Al_2O_3: Cr_2O_3/ZrO_2: HfO_2, *Int. J. High Technol. Ceram.*, **2** (1986) 207–19.

60. Noma, T. and Sawaoka, A. Effect of heat treatment on fracture toughness of alumina–diamond composites sintered at high pressures, *J. Am. Ceram. Soc.*, **68** (2) (1985) C-36, C-17.

61. Orange, G., Fantozzi, G., Trabelsi, R., Homerin, P., Thevenot, F., Leriche, A. and Cambier, F. Mechanical behaviour of zirconia toughened alumina ceramic materials: fracture properties, thermal fatigue and wear resistance. In: *Science of Ceramics*, Vol. 14, Institute of Ceramics, 1987.

62. Orange, G., Fantozzi, G., Homerin, P., Thevenot, F., Leriche, A. and Cambier, F. Thermomechanical properties of zirconia toughened alumina materials: effect of microstructure and temperature on toughening mechanisms. *Zirconia 86*, Somiya, S. (ed.), to be published in *Advances of Ceramics*, The American Ceramic Society, 1986.

63. Noma, T. and Sawaoka, A. Fracture toughness of high-pressure-sintered diamond/silicon nitride composites, *J. Am. Ceram. Soc.*, **68** (10) (1985) C-271, C-273.

64. Tanaka, H., Greil, P. and Petzow, G. Sintering and strength of silicon nitride – silicon carbide composites, *Int. J. High Technol. Ceram.*, **1** (1985) 107–18.

65. Greil, P. and Petzow, G. Sintering and HIPping of silicon nitride–silicon carbide composite materials, *Ceramics International*, **13** (1987) 19–25.

66. McMurtry, C. H., Boecker, W. D. G., Seshadri, S. G., Zanghi, J. S. and

Garnier, J. E. Microstructure and material properties of SiC–TiB₂ particulate composites, *Am. Ceram. Soc. Bull.*, **66** (2) (1987) 325–9.

67. Janney, M. A., Microstructural development and mechanical properties of SiC and of SiC–TiC Composites, *Am. Ceram. Soc. Bull.*, **65** (2) (1986) 357–62.

68. Barnier, P. and Thevenot, F., Synthesis and hot-pressing of single-phase ZrC_xO_y and two phase ZrC_xO_y–ZrO_2 materials, *Int. J. High Technol. Ceram.*, **2** (4) (1986) 291–307.

69   Baik, S. and Becher, P. F. Effect of oxygen contamination on densification of TiB_2, *J. Am. Ceram. Soc.*, **70** (8) (1987) 527–30.

70. Watanabe, T. and Showbu, K. Mechanical properties of hot-pressed TiB_2–ZrO_2 composites, *J. Am. Ceram. Soc.*, **68** (2) (1985) C-34, C-36.

71. Zdaniewski, W. A. Degradation of hot-pressed TiB_2–TiC composite in liquid aluminium, *Am. Ceram. Soc. Bull.*, **65** (10) (1986) 1408–14.

72. Shobu, K. and Watanabe, T. Frictional properties of sintered TiN–TiB_2, and $Ti(CN)$–TiB_2 ceramics at high temperature, *J. Am. Ceram. Soc.*, **70** (5) (1987) C-103, C-104.

73. Stinton, D. P., Caputo, A. J. and Lowden, R. A. Synthesis of fiber-reinforced SiC composites by chemical vapor infiltration, *Am. Ceram. Soc. Bull.*, **65** (2) (1986) 347–50.

74. Lamicq, P. J., Bernhart, G. A., Dauchier, M. M. and Mace, J. G. SiC/SiC composite ceramics, *Am. Ceram. Soc. Bull.*, **65** (2) (1986) 336–8.

75. Bender, B., Shadwell, D., Bulik, C., Incorvati, L. and Lewis, P. Effect of fiber coatings and composite processing on properties of zirconia-based matrix SiC fiber composites, *Am. Ceram. Soc. Bull.*, **65** (2) (1986) 363–9.

76. Fitzer, E. and Gadow, R. Fiber-reinforced silicon carbide, *Am. Ceram. Soc. Bull.*, **65** (2) (1986) 326–35.

77. Mah, T., Mendiratta, M. G., Katz, A. P. and Mazaiyasni, K. S. Recent development in fiber-reinforced high temperature ceramic composites, *Am. Ceram. Soc. Bull.*, **66** (2) (1987) 304–17.

78. Grewe, H., Dreyer, K. and Kolaska, J. Whisker-reinforced ceramics, *Ber. D. K. G.*, **64**, 8/9 (1987) 303–17.

79. Shalek, P. D., Petrovic, J. J., Hurley, G. F. and Gac, F. D. Hot-pressed SiC whisker/Si_3N_4 matrix composites, *Am. Ceram. Soc. Bull.*, **65** (2) (1986) 351–6.

80. Buljan, S. T., Baldowi, J. G. and Huckabee, M. L. Si_3N_4–SiC composites, *Am. Ceram. Soc. Bull.*, **66** (2) (1987) 347–52.

81. Lundberg, R., Kahlman, L. Pompe, R. and Carlsson, R. SiC-whisker-reinforced Si_3N_4 composites, *Am. Ceram. Soc. Bull.*, **66** (2) (1987) 330–3.

82. Homeny, J., Vaugan, W. L. and Ferber, M. K. Processing and mechanical properties of SiC-whisker-Al_2O_3-matrix composites, *Am. Ceram. Soc. Bull.*, **67** (2) (1987) 333–8.

83. Tiegs, T. N. and Becher, P. F. Sintered Al_2O_3–SiC whisker composites, *Am. Ceram. Soc. Bull.*, **66** (2) (1986) 339–42.

84. Chokshi, A. H. and Porter, J. R. Creep deformation of an alumina matrix composite reinforced with silicon carbide whiskers, *J. Am. Ceram. Soc.*, **68** (6) (1985) C-144, C-145.

85. Porter, J. R., Lange, F. F. and Chokshi, A. H. Processing and creep

performance of SiC-whisker-reinforced Al_2O_3, *Am. Ceram. Soc. Bull.*, **66** (2) (1987) 343–7.

86. Tiegs, T. N. and Becher, P. F. Thermal shock behaviour of an alumina-SiC whisker composite, *J. Am. Ceram. Soc.*, **70** (5) (1987) C-109, C-111.

87. Gac, F. D. and Petrovic, J. J. Feasibility of a composite of SiC whiskers in a $MoSi_2$ matrix, *J. Am. Ceram. Soc.*, **68** (8) (1985) C-200, C-201.

88. Marshall, D. B. and Ritter, J. E. Reliability of advanced structural ceramics and ceramic matrix composites. A review, *Am. Ceram. Soc. Bull.*, **66** (2) (1987) 309–17.

89. Kessler, L. W. and Gasiel, T. M. Acoustic microscopy review: nondestructive inspection of advanced ceramic materials, *Adv. Ceram. Mater.*, **2** (2) (1987) 107–9.

90. Friedman, W. D. and Ohnsorg, R. W. The use of X-ray tomography to characterize the effect of HIPing on the density of SiC gas turbine rotors. Communication to the 89th Annual Meeting of The American Ceramic Society Pittsburgh, 26–30 April 1987.

4

Protection Materials: Coatings for Thermal Barrier and Wear Resistance

INGARD KVERNES[a], ERICH LUGSCHEIDER[b] AND YNGVE LINDBLOM[a]

[a]IK Technology AS, Oslo, Norway; [b]Institute of Material Science, Technische Hochschule Aachen, FRG

ABSTRACT

Within coating technology thermal spray techniques are among the most innovative. This paper will discuss the importance of the use of these powder coating techniques, especially in thermal barrier and wear protection, often in combination with corrosion resistance.

The most important progress is made in the field of high energetic procedures, e.g. hypersonic spraying and plasma spray equipment in atmosphere (air plasma spraying — APS), in low pressure (low pressure plasma spraying — LPPS) and under-water plasma spraying (UPS). Robotic control of the plasma spray equipment with up to four simultaneously working powder feeders for the most advanced industrial uses has been applied. The coating materials, e.g. nickel and cobalt base hard alloys, and corrosion resistant alloys, thermal barrier coatings (TBC) and refractory metals, Ta and Mo, are further developed to end qualities, particularly to very fine and pure powders. The microstructure of thermal barrier coatings has been studied and certain requirements are tested. The use and the advantages of TBC in diesel and gas turbine engines are discussed in detail.

Thermal spraying, even for wear protection, covers a wide field of different processes.

Analysis of the complex conditions, such as the characterization of spray materials, mostly powders, powder quality control, process optimization and characterization of the sprayed coatings are gaining more and more importance, especially in high technology applications.

An assessment of the current state-of-the-art and future trends is given.

45

1. INTRODUCTION

The opportunity to improve the surface properties of materials and components with tailor-made coatings has been the biggest achievement of materials technology of the last decade. Surface composites or coated engine parts can now be used in a wide variety of operating conditions, and new technological advantages can be exploited. The most important progress is seen in the field of high energy procedures, e.g. hypersonic spraying and atmospheric plasma spray equipment (air plasma spraying, APS), at low pressure (low pressure plasma spraying, LPPS), and under-water plasma spraying, UPS. Robotic control of the plasma spray equipment with up to four powder feeders working simultaneously for the most advanced industrial uses, has been applied. The coating materials, for example nickel and cobalt based hard alloys, and corrosion resistant alloys, thermal barrier coatings (TBC) and the refractory metals, Ta and Mo, are further developed to give specific final qualities, particularly by using very fine and pure powders. Thermal spraying permits work with high melting TBC materials such as oxides, and wear materials such as ceramics, hard materials, hard metals, hard alloys and the intermetallics.

Ceramic coatings are expected to experience an annual growth rate of 12% through the next decade. By 1995 world sales of ceramic coatings will reach $3·3 billion ($3·3 \times 10^9$) according to Charles H. Kline & Co.[1] The greatest opportunities will be seen in automobile, turbine energy and chemical processing applications. Other forecasters predict the even higher annual growth rates of 15–30%, especially for vapour deposited coatings. The world market for ceramic coatings has been valued at $1·1 billion in 1985. The USA and Western Europe account for 85% of the demand for coatings produced by wet processes, thermal spraying, and chemical and physical vapour deposition. Ceramic coatings are forecast to be introduced in automotive engines by 1995 for cylinder liners, cylinder heads and piston caps. Annual growth is expected to be 25% for engine parts.

2. THERMAL BARRIER COATINGS

In order to put ceramic coatings into regular production a certain level of reproducibility and standard of quality has to be reached. It must be possible to meet a minimum level of consistency with respect to obtainable performance. A review of thermal barrier coatings is given in Ref. 2.

Factors influencing the quality of a plasma sprayed partially stabilized zirconia (PSZ)-coating are:

1. The powder morphology: grain size and shape, density, flowability.
2. The plasma torch: temperature, heat content, velocity, relative movement, spraying distance.
3. The workpiece: temperature, residual stresses, quenching rate of molten particles.

2.1 Powders

Investigating different Y_2O_3-stabilized ZrO_2 coating systems has shown the extreme importance of powder quality. The effects of wide variations in coating performance are thought to be due to chemical purity and homogeneity, in addition to the powder morphology which reflects the manufacturing process. Examples of different powder morphologies are shown in Fig. 1.[3]

Most of the powders used for thermal spraying have sizes between 150 and 5 μm. The ideal powder is one where the grain size range is minimal, and the grain size distribution variation from lot to lot is negligible. Unfortunately, this is not usually the case. The cost of sizeing the powders to narrow limits can, however, often be more than recovered by the improved yield from processing, and from the achievement of more consistent properties in the deposit. The heating characteristics of the powder used in any of the thermal spraying processes are dependent on a variety of factors. The heat transfer from the plasma flame to the material depends on the grain structure and the grain shape factor, and influences the coating in the following manner;

Porous ceramic
Dense ceramic   Increased
Porous sintered metal   heat
Dense sintered metal   transfer
Fused atomized metal   ↓   to the powder

Powder shape: Spherical particles have the lowest surface to volume ratio and will not be heated in the plasma of the combustion torch as quickly as acicular or irregularly shaped particles of the same material and grain size.

Powder manufacturing processes: For example 'spray dried' powders

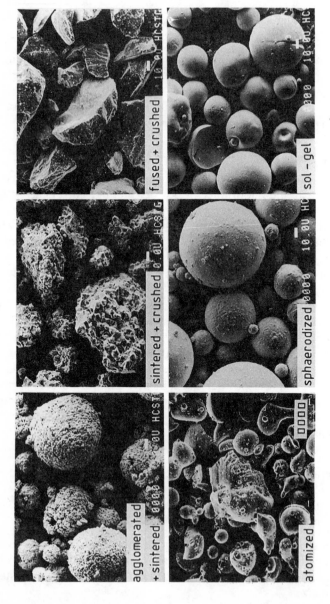

Fig. 1. Different powder morphologies of Y_2O_3 stabilized zirconia.[3]

are manufactured by agglomerating fine powders which have been mixed with a decomposable binder. The blend is then allowed to dry, and the mixture broken, or sprayed into a chamber to form small grains. These powders can vary substantially depending on binder material and content. In contrast, the 'same' material powder size might also be obtained by crushing or grinding from a large solid lump of the material system, forming irregularly shaped powders, or alternatively atomized from a large molten source of the material forming solid (or possibly hollow) spheres. Each of these products will have a different response in spray applications.

For normal thermal spraying operations, the finer powder sizes yield denser coatings. More highly stressed deposits by very finely divided powders tend to oxidize more readily if they are metallic, and also to evaporate in the plasma flame. They also tend to flow with greater difficulty than larger powder sizes, especially metallic powders.

The way that powders are handled and prepared before spraying is too often neglected or under-estimated. The thermal spray powder should be dried and kept clean prior to spraying. The carrier gas used to convey the powder from the feeder to the torch should also be free of oils, water vapour, etc.

2.2 The Plasma Spray Process

Among the most critical spraying parameters are the relative positions of the plasma gun nozzle and the workpiece during the spray operation, and specially the separation distance, and spray angle, as well as the traversing speed at which the gun travels across the surface during spraying (Fig. 2). Optimization of these parameters gives a prescribed microstructure. Changing one of the spray parameters from the optimum, results in a weaker structure. For example, changing the distance from the plasma gun to the workpiece in order to reduce or increase the porosity is not the right way to obtain that result. Selection of powders with the desired manufacturing morphology is the preferable way to change the amount and distribution of porosity in a coating (see powder qualities).

There is no disagreement among coating manufacturers or users as to what is a desired deposition rate. Typical deposition rates that were used for some of the diesel engine parts were 12 μm per pass for the metallic bond coat and 20 μm per pass for the ceramic coating.

With the advent of significant usage of automated thermal spray equipment, it seems that many of the apparently operator-sensitive

Fig. 2. Plasma spray gun.

variations attributed to the operator can be traced to irregularities in either the powder quality used for the coating, or to the storage and handling methods. During the spray operation, the surface preparation has to be varied depending on powder grain size.

2.3 Quality Assurance and Process Control

The composition of each of the coating layers, their thickness, porosity, and other required characteristics should be defined. If we look into a production quality assurance procedure, the following points should be taken into account:

- A visual and dimensional inspection procedure should be carried out.
- A destructive metallographical analysis schedule should be established for both test sample and sample of the part to be sprayed.
- NDT methods are required and are presently under development (Ref. Cost 501, Second Round).
- The quality assurance plan should require complete documentation.

2.4 Structural Features of Zirconia as a Thermal Barrier

Pure zirconia (ZrO_2) is liquid above 2700°C, has a solid cubic structure between 2700 and 2400°C, a tetragonal crystal structure

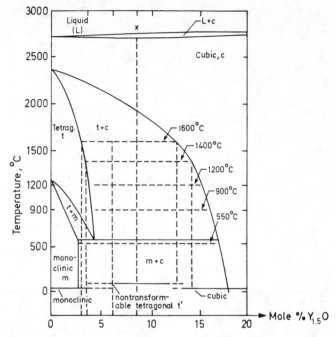

Fig. 3. The ZrO_2 rich section of the $ZrO_2Y_2O_3$ phase diagram.

between 2400 and 1200°C and a monoclinic crystal structure below 1200°C, Fig. 3.[4]

Y_2O_3 and MgO are most frequently used as stabilizers for the desirable tetragonal phase in ZrO_2 in order to suppress the transformation of the tetragonal to the monoclinic structure. One example of an acceptable microstructure is shown in Fig. 4.

I.K. Technology/I.K. Engineering[5] has recently initiated transmission electron microscopy (TEM) studies of plasma sprayed ZrO_2 microstructures. Figure 5 shows a typical TEM image of a 24% MgO stabilized coating. As is expected for materials with high cooling rates (1 MK s^{-1}), as by plasma spraying, the average grain size is small, about 0·4–0·8 μm. This is much less than when sintering zirconia powder. Analyses of the grain boundaries show that they are rich in silicon.

It is interesting to note that the crystalline grain size of the plasma sprayed structure lies in a size range (<1 μm) where the ZrO_2 tetragonal to monoclinic phase transformation is highly grain size sensitive, and

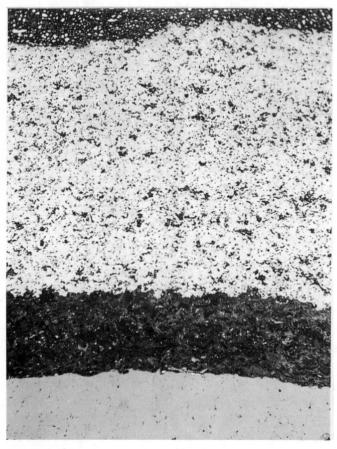

Fig. 4. Microstructure of ZrO₂-24 wt% MgO on FeCrAlY bond layer, as sprayed.

might not take place. It is assumed that this is due to the grain boundary free energies of the tetragonal and monoclinic phases.

In the literature it is shown that the concentration ratio of Y_2O_3 in ZrO_2 is critical. The most spall-resistant ZrO_2 composition is attained with around 8 wt% Y_2O_3 in solid solution homogeneously distributed in the powder particles. The second requirement is to control the residual stresses and porosity during deposition of the powder. The required porosity corresponds to a coating density of $5\cdot2$ Mg m^{-3}. The porosity in the ZrO_2 should increase from the inside of the ZrO_2 to the outer

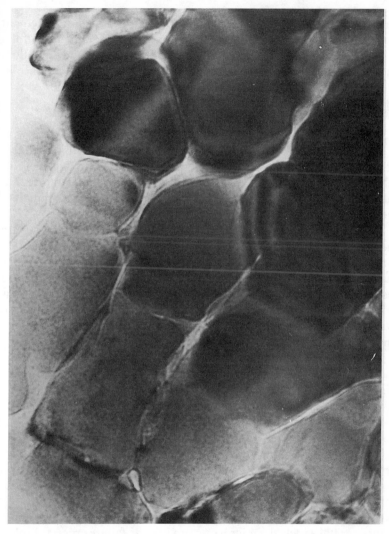

Fig. 5. Bright-field TEM image of sprayed ZrO_2 + 24% MgO, 20 000×.

surface. The third area of progress in the production of high quality ZrO_2 coatings, is the quality control of the powder production method, which has brought greater improvements to the coating quality than many of the other factors. When a bond coat is deposited by low pressure plasma spraying (LPPS), rather than air plasma spraying

(APS), the resistance to spall is doubled. The bond coat improvement from LPPS results from reduced oxidation, and from reduction of the porosity in the metallic bond coat during the deposition, and operation. In comparison, there does not seem to be any significant difference in the spall resistance of Y_2O_3-stabilized ZrO_2 deposited by either APS or LPPS, because no further oxidation can occur.

2.5 Properties

The strength of the ZrO_2 at cycling temperature is definitely dependent on the microstructure, chemical purity and homogeneity, in addition to post-treatments. It is further reported in the literature that the erosion resistance depends substantially on the cohesive strength of the ZrO_2.[6]

Residual stresses in ZrO_2 influence the fatigue strength, fracture toughness and service endurance of the material. A process balanced to form minor comprehensive stresses within the coating is preferable. Such a ZrO_2 coating may exceed 20 000 thermal cycles according to results obtained by NASA. Plasma spraying of ZrO_2 starting with a low substrate temperature may put the ceramic layers near the interphase to the bond coat in tension, and relieve residual stresses through microcracking.

The improvement related to the ZrO_2 tetragonal to monoclinic transformation has been explained in terms of formation of a microcrack zone in front of the propagating crack. These microcracks form due to the strains associated with the tetragonal to monoclinic transformation and toughen the materials by blunting the tip of the microcrack.

Thermal-shock-induced cracking is a common failure mode for ceramics. For heat engine applications, resistance to thermal shock is obviously an important material characteristic. The ranking of ceramics according to their resistance to thermal shock will vary extensively depending on the temperature range of interest, the severity of the thermal shock, the microstructure and the failure criteria. Table 1 shows a comparison between reproducibility of properties obtained in some structural ceramic and plasma sprayed coatings.

2.6 Bond Coats for Thermal Barrier PSZ Coatings

Partially stabilized ZrO_2 (PSZ) has substantial permeability for oxygen and does not protect the substrate from oxidation. When applying thermal barrier coatings on oxidation susceptible substrates

TABLE 1
Representative properties of ZrO_2MgO materials

	Bulk[a] sintered	Coating[b] plasma sprayed
Density (Mg m^{-3})	5·7	4·6–5·2
Melting point (°C)	2 227	2 227
Strength (MPa)	500	35[b]
Young's modulus (G Pa)	200	46[b]
Thermal expansion coefficient (MK^{-1})	9·8	8·0[b]
Thermal conductivity (Wm^{-1} K^{-1})	2·5	0·8[b]
Heat capacity (J kg^{-1} K^{-1})	500	640–670[b]
Fracture toughness (MPa m$^{1/2}$)	4·7	1·26[b]

[a]From I. M. Buckley-Golder: An overview of ceramic applications in engines, The Motor Ship 8th International Marine Propulsion Conference, London, 1986.
[b]Properties obtained by IKT.

an intermediate bond coat is necessary to protect the substrate. The expected lifetime of such a coating system is to a great extent dependent on the bond coat properties.[7, 8]

The bond coats which have been developed are the MCrAlY alloys, where M stands for Fe, Ni, Co or combinations of these. The effects of the elements in the alloys are as follows:

— Chromium: gives oxidation resistance up to 1000°C above which temperature the chromium oxides sublimate.
— Aluminium: gives oxidation resistance above 1000°C. Aluminium can form brittle intermetallic phases in the substrate and therefore the aluminium content has to be kept rather low at 5–10%. At high chromium contents less aluminium is needed to form a protective aluminium oxide film.
— Yttrium: a small content of yttrium increases the service life of the bond coat. Formation of yttrium oxide in the grain boundaries reduces the grain boundary diffusion rate. It has been shown that yttrium is most effective if the yttrium content is not higher than the solubility in the matrix, 0·2%.
— Nickel: to be preferred on nickel base substrates.
— Cobalt: preferably on cobalt based substrates. Cobalt makes the

bond coat harder and less ductile than nickel, but gives a bond
coat more resistance against sulphidation.
— Iron has the best hot corrosion resistance properties. If used on
nickel base alloys there are difficulties with diffusion of nickel into
the bond coat, changing the crystal structure from body centre
cubic to face centre cubic.

The quality requirements for bond coats are the following:

— Oxides: from the viewpoint of bond coat ductility, the oxide
content should be low. Oxides with proper morphology can,
however, be obstacles to the diffusion of Ni into the bond coat and
retard nickel diffusion by a factor of 25.[9] The oxide content
increases with increasing heat input.
— Porosity: should be as low as possible, and less than 3%. Porosity
increases the permeability of oxygen and sulphur through the
bond coat.
— Mass gain: the mass gain during thermal cycling of the component
indicates the rate of degradation of the bond coat–thermal barrier
system. The mass gain has a strong relationship to heat input
during spraying, Fig. 6.[7]

3. CERAMIC INSULATED ENGINE PARTS

3.1 Diesel Engine

The present trends in engine technology are to increase engine
ratings, reduce weight and fuel consumption and generally raise the
efficiency. Such an engine is, for example, a low heat rejection (LHR)
diesel engine in which the main requirement is the reduction of heat
loss to the engine block cooling system with a consequent increase in
available energy in the exhaust gas. Thermal insulation of the diesel
engine combustion chamber components reduces heat transfer to the
coolant, but these hot surfaces also heat the intake air, thereby reducing
the number of air molecules admitted before inlet valve closure and the
start of compression, resulting in a lower compression pressure. This is
referred to as the volumetric efficiency penalty, which would result in
higher fuel consumption with naturally aspirated engines. Thus the
increased energy in the exhaust gas must be used so that it compensates
for the volumetric efficiency loss, as well as providing an overall positive
gain. Turbocharging with the increased exhaust energy resulting from

	Bond coating		Thermal barrier coating	
	A	V	A	V
●	300	26	500	31
○	300	26	550	32
□	350	27	500	31
■	350	27	550	32
▽	350	27	600	33
▼	400	28	550	32
△	400	28	600	33
▲	400	28	700	35
×	450	29	650	34
+	450	29	700	35
*	500	30	700	35

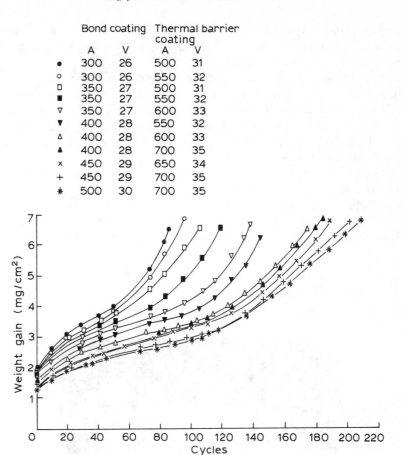

Fig. 6. Effect of heat input during spraying on weight gain for a Ni-16, 8Cr-5·8Al-0·31Y/ZrO_2-7·9Y_2O_3 during thermal cycling 820–1095° C (cycle = 6 min heating to highest temperature, holding 60 min at highest temperature, 60 min cooling to lowest temperature).[7]

thermal insulation of the combustion chamber produces a higher boost pressure for the same engine setting. Several companies are developing a second turbine in the exhaust, either before or after the turbocharger, which is mechanically connected to the output shaft. This arrangement is referred to as turbocompounding and it improves overall efficiency as well as increasing output power.

The engine should be adjusted to the changed working conditions.

Compression, injection parameters and turbocharger should be adjusted to maintain the specified maximum cylinder pressure, and minimizing fuel consumption.

The possibility especially for better combustion of low grade heavy fuel is important in such research effort. The one fact alone that ceramic coatings have been able to reduce the specific fuel consumption and increase the energy content of the combustion gases gives indications of the possibility for reducing operational costs, to more than compensate for coating costs. As an additional advantage there are the already documented reductions of thermal load on the components, which improves service life, particularly where the combustion bowl design has thin surface sections.

The following parts of the diesel engine hot-section provide insulation advantages with thermal barrier coatings of the stabilized ZrO_2 type:

- piston crown (Fig. 7)
- exhaust valves (Fig. 8)
- cylinder head (firedeck)

Fig. 7. Piston crown coated with ZrO_2-MgO + FeCrAlY.

Fig. 8. Exhaust valve coated wth ZrO_2-MgO + FeCrAlY.

- exhaust port*
- inlet air passage*
- exhaust gas passage inlet to turbocharger
- cylinder liner from top ring reversal point to head.

These coatings should be engineered to the components. Thickness should be determined analytically for specific engines. Injection parameters should be adjusted for the higher temperatures. Turbochargers should be rematched for the higher energy levels resulting from insulation. Variable geometry turbochargers appear better suited to the wider range of energy levels. Turbocompounding should be considered. Other waste heat utilization schemes are in development, but have a high capital cost, are complex and increase maintenance requirements. However, higher fuel costs could make some of these systems cost-effective.

Tests should be made with a real engine because transient conditions in the cylinder will be affected by the coatings. The coatings should be PSZ *with* surface treatment. Figure 9 shows the Cummins TACOM

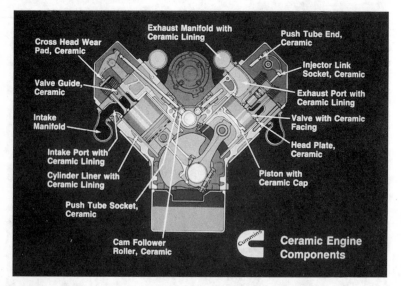

Fig. 9. Ceramic insulated engine components and potential ceramic wear components.

*Major effort with cast in place inserts of aluminium titanate.

ceramic insulated engine components, and potential ceramic wear components.

3.2 Gas Turbine

Thermal barrier coatings have been used on combustor linings, transition ducts, and vane platforms in production aircraft gas turbine engines for at least 17 years. Application of thermal barrier coatings to turbine airfoils provides the major advantage of allowing an increase in the turbine inlet temperature to 1500°C with the accompanying improvement in fuel consumption or extended life at current firing rates. However, thermal barrier coatings are not yet qualified for production use on turbine airfoils. Thin ceramic coatings have demonstrated outstanding hot-corrosion resistance on turbine airfoils. Thinner ceramic coatings have outstanding hot-corrosion resistance on turbine airfoils, and have a greater ability to accommodate thermal mismatch stresses. Figure 10 shows a cross-section of a gas turbine engine indicating the coated parts.

4. WEAR RESISTANT COATINGS

The importance of thermal coating techniques especially in wear protection, often combined with corrosion resistance, is determined by

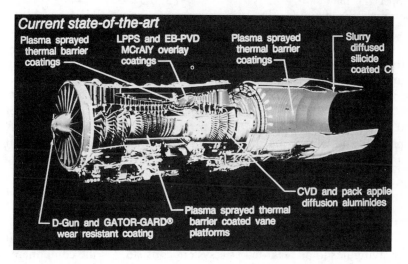

Fig. 10. Gas turbine engine coating usage (Pratt & Whitney).

TABLE 2
Material types and thermal spray processes for wear protection

Wear coating material types	Typical coating processes

Hard alloys:
  self-fluxing Fe-, Co-, Ni base
  alloys
	Flame spraying (FS)
	Air Plasma Spraying (APS)

Oxides based on:
  Al_2O_3
  TiO_2
  Cr_2O_3
  ZrO_2
  Cermets (e.g. Cr_2O_3/NiCr)
	Air Plasma Spraying (APS)
	Detonation-gun (DG)
	Jet-Kote (JK)
	Vacuum-Plasma-Spraying (VPS)
	Inert-Plasma-Spraying (IPS)
	Underwater-Plasma-Spraying (UPS)

Hard metals:
  WC-Co
  Cr_3C_2-NiCr
	APS, DG, JK, VPS, UPS, IPS

Hard materials	

Refractory
  carbides   WC, TiC
  nitrides   ZrN, TiN
  carbonitrides Ti(CN)
  borides TiB_2
  silicides $MoSi_2$
	VPS, IPS (APS)

Special ceramics: Si_3N_4, BN, B_4C, SiC, AlN	VPS, IPS (APS)

the great damage caused by wear and corrosion — in West Germany and Norway approximatedly 5% of the gross national product, which means about 75 billion DM for Germany and 7 billion DM for Norway.

Thermal spraying even for wear protection covers a wide field of different coating processes (Table 2): the reactive gas processes such as powder and wire flame spraying; the high velocity processes D-Gun and Jet Kote; wire-spraying; and of course the different high energy plasma spray processes. Laser spraying is just under development.

At the present time the improvement of coating properties, even of wear coatings by special post treatments such as LASER, HIP and VED (Vacuum Explosion Densification), are topics of extensive research.

In order to obtain successful and economic coatings even thermal spraying has to be regarded as a system consisting of the components substrate, coating material and coating process. This system exists for every engineering surface that has to be coated, and all components of the system and their interactions have to be optimized. Analysis of the complex service conditions, the characterization of the spray materials, which are mostly powders, powder quality control, process optimization, and characterization of the sprayed coatings, are gaining more and more importance, especially in high technology applications.[10]

4.1 Wear Coating Material Types for Thermal Spraying

Great advances in spray materials combined with progress made in spray systems, process developments and automation, have widened the range of wear materials, essentially in the last decade.

Typical coating material types and thermal spray processes are shown in Fig. 11. These material ranges will be widened with regard to wear applications combined with protection against corrosion, hot gas corrosion, and thermal and electrical insulation, by coating materials such as MCrAlY, stabilized ZrO_2, intermetallics, superalloys, refractory metals, and high alloyed steels, which are already in use.

For metal-base coating systems such as the important self-fluxing wear resistant Ni-Cr-B-Si hard alloys, mainly the lower energy flame-spraying is the economic coating process. The high energy plasma

Fig. 11. Flame spraying of Ni-Cr-B-Si wear coatings using oxygen-acetylene burner.

processes are the most important for processing the high melting wear materials of the oxide and hard metal families. Air plasma sprayed oxides and hard metal coatings are well established in almost all industries for high degrees of wear protection. For WC-Co and Cr_3C_2-NiCr coatings use of the high velocity flame spray process, Jet Kote, is increasing, especially in the USA.

The most advanced vacuum plasma spraying is an important alternative for conventional coating materials if extreme coating quality is required. Vacuum plasma spraying has already made it possible to spray extremely oxygen sensitive powders, such as the refractory metals tantalum and titanium, mainly for corrosion protection,[11] and materials sensitive to oxidation and decomposition, such as the carbides and other refractory hard materials. The development of high quality coatings of for example TiC, Ti(CN), and TiB_2 by vacuum plasma spraying in the last few years offers further new applications in the field of wear protection.

At the present time the development of high technology ceramics such as BC, Si_3N_4, SiC, BN and AlN is a topic of extensive research, in which the peritectic decomposition of such materials during the spray process is the main problem to be solved.

4.2 Metallic Wear Systems

For metallic wear systems very often welding processes are the most important coating processes, for example, for cobalt base hard alloys of the Stellite type.[12] Within the numerous metal base wear coatings used in industry, self-fluxing hard alloys on a nickel base should be mentioned in this paper because of their importance for wear protection.[12] In Western Europe about 1000 t year^{-1} of Ni-Cr-B-Si coating powder are used for flame spraying of wear resistant coatings. Nickel-chromium-boron-silicon hard alloys mainly produced by gas atomizing, consist of a Ni-Cr matrix with embedded hard phases of nickel and chromium borides, silicides and carbides. The amount and the type of the wear resistant phase depends on the alloy composition, and the numerous grades used in industry can be classified by their HRC hardness in 20, 40, 50 and 60 type alloys shown in Table 3.

Ni-Cr-B-Si alloys are normally processed by a so-called two-step process. The flame sprayed coatings (Fig. 11) are heat treated by a second step between their solidus and liquidus temperature (1000–1080° C). This leads to a high coating densification (Fig. 12) and an essential increase of the bond strength to the base material, and can be

TABLE 3
Wear resistant Ni-Cr-B-Si hard surfacing coatings

Alloy type HRC	Chemical composition (wt %, bal. nickel)					Wear resistance	Oxidation resistance	Impact sensitivity	Hot hardness
	B	Si	C	Cr	Fe				
20	1–1.25	3–4	0.25	5	2–3.5	5	2	1	1
40	1.5–2.25	2–3	0.5	7.5–10	2.5–3	4	2	2	1
50	2–2.5	3–4	0.5–0.7	10–13	3.5–4.5	3	2	3	1
60	3–3.5	3–4,5	0.5–0.8	13–17	4–4.75	1	1	4	3

N.B. 1 = optimum; 5 = very poor.

Fig. 12. Micrograph of a flame sprayed Ni-Cr-B-Si wear coating of type 60 (M = 200×).

Fig. 13. Induction treatment of a flame sprayed Ni-Cr-B-Si wear resistant coating.[11]

TABLE 4

Industrial applications of wear resistant Ni-Cr-B-Si coatings[13]

Partial/large area coatings	Steel/machine constructions	Mining	Concrete	Wood	Ceramic	Sewage, silt	Armatures	Glass industry
Fans	×	×	×	×		×		
Chutes	×	×						
Grinding balls					×			
Guides	×	×						
Strippers		×						×
Wear plates	×	×	×		×	×		
Centrifuges		×				×		
Mixers	×	×	×		×	×		
Seat rings	×						×	
Valves	×	×				×	×	×
Moulds	×							×

carried out using a special oxygen-acetylene burner, vacuum furnace treatment, or induction heating. For large scale production, vacuum furnace melting is preferred (1 mPa). High quality coatings, and a high reproducibility of coating quality, can be obtained by induction heating, in general used for cylinders and tubes up to diameters of 2 m (Fig. 13).

Typical applications of Ni-Cr-B-Si wear coatings are shown in Table 4.

In the turbine industry blades are coated by air pressure spraying (APS) followed by a vacuum furnace heat treatment. High chromium containing grades are preferred in these applications.[13]

An essential increase of the wear resistance of Ni-Cr-B-Si coatings, especially for abrasive wear stress, can be obtained by spraying powder mixtures containing up to about 50% of refractory carbides such as $WC(WC-W_2C)$, NiC VC, NbC, TaC. For spraying these so-called pseudo-alloys the process must be optimized in such a way that the carbides are not dissolved in the Ni-Cr-B-Si matrix, but remain strongly bonded to the matrix by diffusion.

Recent developments have led to a new class of self-fluxing nickel base coating alloys with high wear resistance and a higher corrosion resistance in comparison to conventional Ni-Cr-B-Si coatings. The wear resistance of the so-called τ-alloys is given by complex τ-borides of the composition $Ni_{21}B_6X_2$, where X is a refractory metal such as Ta, Nb, or Ti, stabilizing the hard phase structure within the coating alloy.[14]

4.3 Oxide Ceramic Wear System

Plasma spray processes are characterized by the highest energy input into the coating powders during spraying, and are therefore especially appropriate for spraying high melting oxides for wear protection. Air plasma spraying (APS) especially has gained a dominant economical role in spraying wear resistant oxides in many industries, Table 5. In Table 6, typical oxide grades for wear applications used successfully in industry are listed. Using high velocity plasma spraying, high quality oxide coatings can be produced containing less than 2% of closed porosity, Fig. 14. To improve the bond strength, bond coatings consisting of for example NiAl, NiCr and NiCrAl are often sprayed. Typical bond strengths of ceramic coatings are in the range of 50–70 MPa. The surface roughness of APS coatings after spraying of less than 20–30 μm (R_t) can be lowered by grinding with diamant disks to less than 0·7 μm (R_t). Typical grinding data for ceramic coatings in wear applications are given in Table 7.

TABLE 5
Typical industrial applications for
plasma sprayed oxide wear coatings

Textile industry
Synthetic fibres
Pneumatic hydraulic systems
Paper-printing industry
Turbines
Pumps
Plastics industry
Marine industry
Off-shore industry
Steel industry
Automotive industry
Mechanical engineering
Chemical engineering
Electronics—computers

TABLE 6
Typical wear resistant oxide grades for thermal spraying

Al_2O_3
Al_2O_3-$X$$TiO_2$ (X = 3, 13, 30, 40)
Al_2O_3–$X$$Cr_2O_3$ (X = 2, 27, 50)
Al_2O_3-$X$$ZrO_2$ (X = 25, 40)
Mullite ($3Al_2O_3 . 2 SiO_2$)
Magnesium aluminium spinel (Mg Al_2O_4)
TiO_2
Barium titanate ($BaTiO_3$)
Cr_2O_3
Cr_2O_3-$X$$TiO_2$ (X = 3, 40, 50, 55)
Cr_2O_3-$3TiO_2$-$5SiO_2$
ZrO_2-stab. (CaO, MgO, Y_2O_3)
Zircon ($ZrSiO_4$)
Magnesium zirconate ($MgZrO_3$)

TABLE 7
Grinding wheel, for example type Al-450-20-4-K400N, D46-C75

	Flat surface grinding	Roll grinding
Peripheral speed (m/s)	30	35/50
Table feed (m/min)	25	—
Infeed lift (mm)	0·01	0·01
Cross feed (mm)	10	10

Fig. 14. Microstructure of an APS sprayed Al_2O_3-13TiO_2 wear coating. Bond coat NiCr; M = 200×.

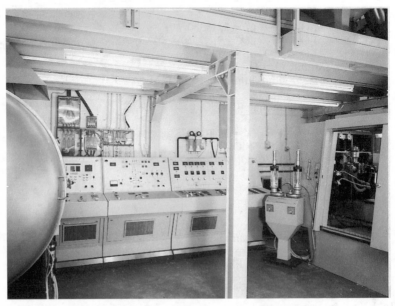

Fig. 15. Most advanced air plasma spray system (Material Science Institute, Technical University Aachen).

Fig. 16. Underwater plasma spraying of wear resistant coatings (Material Science Institute, Technical University Aachen).

Fig. 17. Micrograph of an UPS coating consisting of Al_2O_3 and a NiCr bond coat (M = 200×).

Today the most advanced automatic plasma spray equipment and handling systems are available for the efficient coating of large scale production runs and complicated individual components (Fig. 15). For spraying high quality and reproducible coatings programmable industrial robot systems have been successfully used.[10]

Recently a new plasma spray process is under development at Material Science Institute of Aachen University, whereby the coating process is run under water, Fig. 16.[14] Besides advantages such as the reduction of noise, dust pollution and radiation, the so-called under water plasma spraying (UPS) system leads to an essential improvement of coating quality, especially when UPS is done under high pressure water in specially designed pressure vessel systems. Many coating materials have been tested by UPS, and even ceramic materials lead to high quality coatings, Fig. 17. Applications for UPS can be seen in conventional wear protection, as well as in marine and offshore industries.

4.4 Hard metals for wear applications

Wear resistant hard metals consisting of a matrix of a metallic hard material, mainly a refractory carbide, and a metal binder, well known from the cemented carbide industry, are very successfully used for wear protection by thermal spraying. Typical WC-Co grades contain 6, 12, 13, 17 and 20 wt-% of cobalt. Other hard-metal type materials are $93Cr_3C_2$-$7NiCr$, $75Cr_3C_2$-$25NiCr$, and $83Cr_3C_2$-$17Ni$.

For conventional applications APS is used for producing high quality coatings.[16] In aircraft and the turbine industry vacuum-plasma spraying and D-gun coatings result in hard metal coatings with extremely high bond strength, Fig. 18.

Typical applications for wear resistant hard metals are for example, coatings for bearings in steam and gas turbines, centrifugal pumps and turbo-compressors, nuclear armatures, and end supports for blades in stationary and jet engine turbines.[17] For applications where wear and additional corrosion resistance is required instead of WC-13Co, coatings of WC-20Cr-7Ni and WC-10Co-4Cr are preferred.

As mentioned before, especially for conventional wear protection with WC-Co and Cr_3C_2-NiCr type powders, the use of the Jet Kote process is increasing.[18] This hypersonic flame spray process, quite new in industry, provides essential higher deposition rates, in comparison with other spray techniques, Fig. 19.

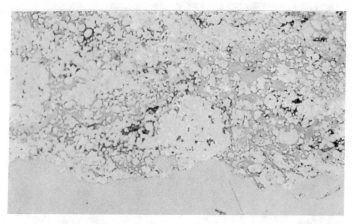

Fig. 18. Micrograph of a VPS WC-12Co wear resistant coating (M = 500×).

4.5 Refractory Hard Material for Wear Protection

In the family of the carbides, nitrides, borides and silicides of the IVA, VA and VIA groups of the periodic system of the elements (Ti, Zr, Hf/V, Nb, Ta/Cr, Mo, W) many refractory hard materials with outstanding properties exist; for example TaC with the highest melting point of about 4000 °C, TiC and TiB$_2$ with almost 3000 Vickers hardness, MoSi$_2$ with an oxidation resistance up to 1600° C.[19] This is the reason why these materials were always under consideration as potentially useful coating materials for thermal spraying. The sensitivity of most of these components to oxidation and decomposition during spraying, decarburization of carbides for example, has hindered the application of these interesting materials by the air spray processes. Some success has been obtained in the past using shrouded inert gas plasma spraying (IPS), but there has now started extensive development of refractory hard material coatings, since highly advanced vacuum plasma spray equipment has become available in industry, resulting in high quality VPS coatings for wear protection.

Vacuum plasma spraying (VPS) (i.e. spraying in a chamber at reduced argon pressure (5–20 kPa)) offers important advantages in comparison to other thermal spray processes, Fig. 20,[21] which are:

- Higher particle velocity, resulting in denser coatings and higher bond strength.

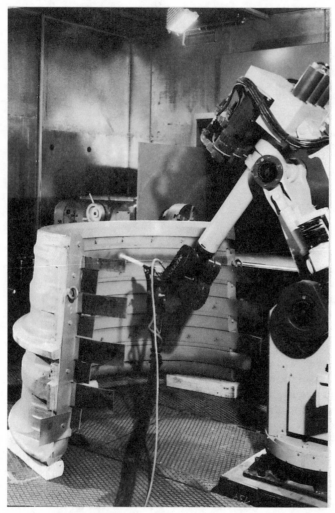

Fig. 19. Jet Kote spraying of wear resistant Cr_3C_2-NiCr coatings.[20]

- Substrate heating is possible because of the inert vacuum atmosphere, resulting in an important increase of diffusion between coating and substrate material and therefore an improvement of bond strength.
- High purity, no oxygen reaction, high phase stability of the coatings, because of the inert gas vacuum process.

Advanced VPS production equipment consists of three modules, which

Fig. 20. Most advanced vacuum plasma spray system (Material Science Institute, Technical University Aachen).

Fig. 21. VPS TiC-NiCr coating for wear protection (M = 500×).

can be controlled individually as one unit in either the manual or the automatic modes. The vacuum system is composed of a vacuum chamber, filter and vacuum pumping unit (final pressure 0·1 Pa) with its own control console for adjustment and control of the inert gas atmosphere in the range of 0·05 to 20 kPa. The plasma spraying unit consists of a plasma torch, powder feeder and plasma control panel for adjusting and stabilizing the current operating parameters. The powder feeder consists of two powder containers, which could be operated individually or both at the same time. An exactly reproducible powder flow rate is guaranteed even for powders with poor flow properties, due to the installation of a stirrer and a heater for drying the powders before spraying. The powder containers are vacuum sealed, which allows for degassing of the spraying powders under vacuum or in an inert gas atmosphere. The handling system is controlled by a programmable CNS system with five axes, or programmable industrial robot systems are used.

With the VPS process during the last few years high quality coatings of for example, TiC, Ti(CN) and TiN, for wear applications have been developed, as well as TiB_2 coatings for, for example, applications where components have to be protected against molten metals. The quality of oxidation-resistant $MoSi_2$ and WSi_2 coatings could be improved significantly using VPS. Vacuum plasma spraying also enables the production of advanced refractory hard metals — metal coatings such as TiC-NiCr, and TiB_2-Ti, Fig. 21, and offers advantages concerning coating quality in spraying well-known and widely used industrial coatings such as WC-Co and Cr_3C_2-NiCr.

4.6 Improvement of Thermal Spray Coatings by Post-treatment

The properties of thermal spray coatings can be modified or improved by LASER treatment,[22/23] LASER alloying, cladding and glazing in combination with thermal spraying, which broadens the area of applications for sprayed coatings. Increasing density, bond strength, improved microstructure, and metallic coatings with an almost amorphous structure at the surface, are important advantages.

Figure 22 shows a LASER treated APS sprayed nickelbase hard-surfacing coating, resulting in an increase of wear resistance against adhesion wear stress of 60%.

Hot isostatic pressing (hipping) is also used for the treatment of thermal sprayed coatings to obtain denser coatings with extreme adhesion to the substrate. The structure of a hipped APS coating of a

Fig. 22. LASER remelted APS hardsurfacing coating (M = 200×).

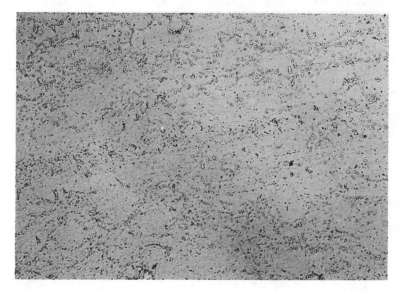

Fig. 23. HIP treated APS hardsurfacing coating[16] (M = 200×).

hard surfacing alloy is shown in Fig. 23. For large sample components and internal coatings densification using the vacuum explosion densification process may offer improvements similar to those of hipping.[24]

ACKNOWLEDGEMENT

The authors are grateful to H. Lønstad for support in preparing and editing the manuscript.

REFERENCES

1. Kline, C., *Advanced materials technologies, Report 7, Ceramic coatings*, October 1986.
2. Bürgel, R. and Kvernes, I., *Thermal barrier coatings*, Proceedings of a conference, Liege 6–9 October 1986, Reidel Publishing Company.
3. Kvernes, I. and Hoel, R. H., *Advanced coating developments for internal combustion engine parts*, SAE 870160, Febr. 1987.
4. Miller, R. A., Smialek, J. L. and Garlick, R. G., Phase stability in plasma sprayed partially stabilized Zirconia-Yttria, *Advances in Ceramics, Vol. 3, Science and Technology of Zirconia*, The American Ceramic Society Inc., Columbus, Ohio, 1981, p. 241.
5. Hoel, R. H. and Kvernes, I., *The microstructure of thermal barrier coatings*, Conference 14–17 September 1987, Orlando, Florida, ASM.
6. Elssner, L. and Kvernes, I., *Mechanical characterization of ZrO_2 coatings*, Proceedings US Navy NATO Advanced Workshop, Aquafredda, Italy, April 1984.
7. Stecura, S., *Effects of plasma spray parameters on two layer thermal barrier coating system life*, NASA TM 81724, March 1981.
8. Stecura, S., Effects of yttrium, aluminium and chromium concentrations in bond coatings on the performance of zirconia-yttria thermal barriers. *Thin Solid Films,* **73** (1980) 481–9.
9. Burman, C., Ericsson, T., Kvernes, I. and Lindblom, Y., *Coatings with lenticular oxides preventing interdiffusion*, International Conference on Metallurgical Coatings, San Diego, CA, USA, 23–27 March, 1987.
10. Nicoll, A., *Production thermal spray equipments and quality control consideration*, Plasma-Technik AG, Technical Publication No. 8600E, Switzerland.
11. Reimann, H., *Induction Fusing*, Gotek GmbH, Technical Publication No. 3073/5.81, Frankfurt/M, FRG.
12. Knotek, O., Lugscheider, E. and Eschnauer, H., *Hartlegierungen zum Verschleiss-schutz*, Düsseldorf, Verlag Stahleisen, 1975.
13. Villat, M., *Beschichtungen durch thermisches Beschichten*, SGT-Fachtagung, 1985, Bern.

14. Lugscheider, E., Krautwald, A., Eschnauer, H. and Meinhardt, H., *A new type of atomized coating powder for protection against wear and corrosion*, International Conference on Metallurgical Coatings, San Diego, CA, USA, Elsevier Sequoia SA, Lausanne and New York, 1987.
15. Reimann, H. Gotek GmbH, Technical Publication No. 3122 d/4.87, Frankfurt/M, FRG.
16. Chandler, P. E. and Nicoll, A., *Plasma sprayed tungsten carbide cobalt coatings*, 2nd International Conference on Surface Engineering, Stratford-upon-Avon, UK, 1987.
17. Lugscheider, E., Beschichtungen für den Hochtechnologie-bereich, *Metallo-berfläche,* **41** (1987) 240.
18. Kreye, H., International Thermal Spray Conference, 1986, Montreal, Canada Pergamon Press.
19 Eschnauer, H. and Lugscheider, E., *Metallic and ceramic powders for vacuum plasma spraying*, International Conference on Metallurgical Coatings, San Diego, CA, USA, Elsevier Sequoia SA, Lausanne and New York, 1984.
20. Gotek GmbH, Frankfurt/M, FRG, unpublished.
21. Lugscheider, E., *Vakuumplasmaspritzen — Ein modernes Beschichtungs-verfahren*, Battelle Kolloquium, Oberfläche, 1985, Frankfurt/M, FRG.
22. Lugscheider, E., Krautwald, A. and Wilden, J., *Laser treatment of thermal spray coatings*, Conference on Heat Treatment, 1987, Institute of Metals, London, UK.
23. Hoel, R. H., Kvernes, I. and Lugscheider, E., Overflatebehandling med laser, *Teknisk Ukeblad (Norway),* **16** (1987), 56.
24. Hampel, H., Lugscheider, E. and Lehrheuer, W., Explosivfügen metallischer Werkstoffe, *Die Umschau,* **12** (1986) 617.

5

The Production of High-Grade Technical Ceramics

B. G. NEWLAND

Morgan Matroc Ltd, Stourport-on-Severn, Worcs, UK

ABSTRACT

Improvements in production technology are important for increasing the use of high-grade technical ceramics. Materials, process methods and equipment which derive from traditional practices or related industries are being replaced by an integrated approach to zero defect manufacture. The benefits are demonstrated in terms of microstructural control and superior properties. Such achievements further enhance the desirability of technical ceramics, and consequent growth in demand will increase the cost effectiveness of the new technology.

1. INTRODUCTION

The growing interest in 'high-tech' ceramics arises from their basic material structure, that is, strong chemical bonds and the mixed ionic-covalent nature of these bonds. This bonding is responsible for the characteristic high hardness, high temperature strength and relative inertness of many technical ceramics; but it can also produce both electrically insulating and semiconducting behaviour, optical transparency, ferromagnetic and ferroelectric (piezoelectric, pyroelectric) behaviour. The growing use of 'high-tech' ceramics is based also on their ability to combine both functional and structural rôles in one operation.

Ceramic production technology is the key to the actual fulfilment of the wide ranging potential of technical ceramic materials, with better performance and reliability coming from improvements in the science and engineering of ceramic processing.

The production of technical ceramics is currently a derivative technology, that is it uses raw materials and processes derived from those developed initially for the minerals, metals, chemicals and pottery industries. The skill of the technical ceramics producer lies in adopting and adapting these processes from different technologies, but this procedure is the source of many of the technical and economic problems which must be overcome to ensure further significant expansion in the applications of technical ceramics. It also limits the processing of large ceramic parts and assemblies.

The production technology of technical ceramics, however, is at the threshold of a radical change — from being an assembly of derivative materials and processes it could become a fully integrated process cycle using specific starting materials.

2. THE PRODUCTION ROUTE

Most technical ceramics are manufactured by a sequential production route similar to that shown in Fig. 1. The options available at each stage facilitate the production of a wide range of materials and components.

The traditional production route used natural raw materials, for example clay and talc, which were formed into relatively simple shapes by 'damp' compaction, for example extrusion or pressing, and then dried and liquid phase sintered. The resultant ceramics were suitable for relatively low performance applications, electrical insulators for example, but they have since been largely replaced by plastics which can be produced by more cost effective technology.

Present day production routes combine technologies derived from the powder metals, polymers and traditional ceramics industries. These have been adapted and developed to the stage where technical ceramic components can be produced in large quantities for cost effective use in high performance applications. Nonetheless, most existing commercial ceramic powders are produced from minerals by conventional mining, beneficiation and extraction techniques. This makes it difficult to control the particle shape and size distribution, and impurity content, and is a major factor limiting the reliability of current technical ceramics. Furthermore, many of the processes currently used for the

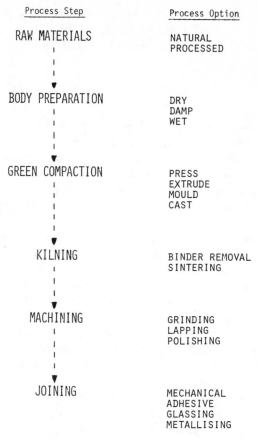

Fig. 1. Production routes for technical ceramics.

production of technical ceramics originate from industries where the powder characteristics which determine good compaction and sintering are different from those required to produce reliable, high quality technical ceramics. In addition, the post-sinter machining of technical ceramics and their joining to other materials is more difficult to achieve using existing metals-orientated technologies because of the hardness, brittleness and inertness of ceramics.

Therefore, further major improvements in the production route are being vigorously investigated. The major objectives are to increase the reliability of technical ceramics by reducing the size and number of impurities and defects, and to lessen the need for post-sintering machining by improving near-net-shaping techniques for complex parts. Significant developments are underway at all stages.

3. POWDERS

The use of 'sub-micrometre' powders has been widely promoted for the improvement of ceramic properties. The results of using such powders to produce a sintered 99·7 mass % alumina ceramic are shown in Table 1; a disappointing increase in flexural strength is obtained. The microstructure of the 'improved' ceramic (Fig. 2) still contains pores and large grains which act as critical flaws in limiting the flexural strength.[1]

Ceramic powders are usually classified as sub-micrometre by measurement of the average (d_{50}) particle diameter (Fig. 3). It is known, however, that ceramic powders contain agglomerates which greatly affect the evolution of microstructure.[2] These agglomerates are made up of smaller and more densely packed sub-units,[3] which have been called domains.[4] The agglomerates and domains can be described as either 'hard', that is cemented or partially sintered groups of particles, or 'soft', that is groups held together by weak Van der Waals' forces. The relative proportions and sizes of the hard and soft agglomerates are determined by all the stages involved in producing the ceramic powder. The 'sub-micron' powder used to produce the 'improved' 99·7 mass % alumina ceramic contained a significant proportion of hard agglomerates/ domains whose average size was much greater than 1 μm. They are difficult to break down in subsequent processing.

Therefore, care must be taken to select a powder containing the smallest possible fraction of hard agglomerates/domains. The benefits thus obtained can be demonstrated in the production of sintered 99·9

TABLE 1
Effect of powder particle size (d_{50}) on strength of 99·7 mass % alumina ceramic

Alumina content (mass %)	99·7	99·7
Powder particle size, d_{50} (μm)	1·2	0·6
Ceramic mean grain size (μm)	8	3
Ceramic flexural strength, MPa	330	380

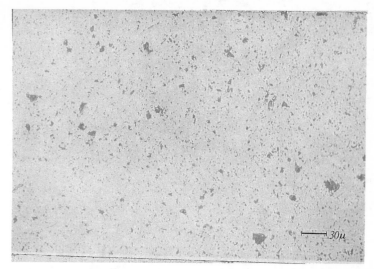

Fig. 2. Extensive closed porosity in the polished surface of an 'improved' 99·7 mass % alumina ceramic.

mass % alumina ceramics (Table 2). Both powders are commercially available, but that with the smaller d_{99} particle size produces a very high performance ceramic. The resultant microstructure (Fig. 4) exhibits minimal porosity and fine even grain size, which contribute to improved strength and reliability.

Recent developments have resulted in the production of true submicron powders containing only soft agglomerates. They are produced by chemical reaction, nucleation and growth from the gaseous or liquid phase where the interparticulate attraction arises only from surface energy effects or absorbed layers. However, subsequent heat treatment to increase purity can result in the formation of hard agglomerates.

4. GREEN COMPACTION

The ideal powder for producing a dense sintered ceramic microstructure should preferably consist of agglomerates which are strong enough to flow easily and loose pack uniformly, but are also soft enough to be broken down into domains and individual particles by relatively small applied consolidation forces. The resulting unsintered ('green') microstructure should then consist of a large number of small domains with

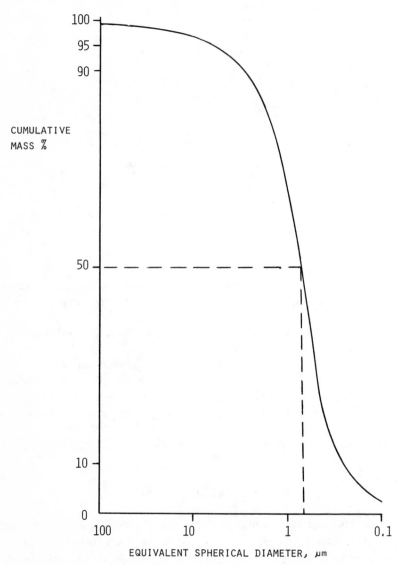

Fig. 3. Particle size distribution of a typical ceramic powder.

TABLE 2
Effect of powder particle size (d_{99}) on strength of 99·9 mass %
alumina ceramic

Alumina content (mass %)	99·9	99·9
Powder particle size, d_{99} (μm)	20	5
Ceramic density (g/cm^3)	3·95	3·97
Ceramic max. grain size (μm)	6	3
Ceramic mean grain size (μm)	3	<2
Ceramic flexural strength, MPa	400	550

fine, evenly distributed inter-domain porosity. At present it is difficult to obtain particles with these characteristics; control of the interparticulate forces is poor even with the new chemically produced powders.

Although sub-micrometre powders are desirable for the production of high grade technical ceramics, they are difficult to process using the conventional dry powder pressing technology. Therefore, it is necessary to produce powder agglomerates, but these are usually made hard by the addition of processing aids (binders, lubricants). While such a press 'body' is suitable for volume production of components, it gives green microstructures which are non-uniform and include contaminants from the powder processing. Subsequent sintering of these green

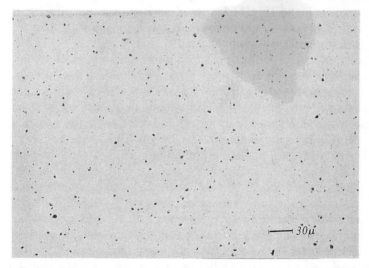

Fig. 4. Reduced closed porosity in the polished surface of a high-grade 99·9 mass % alumina ceramic.

KEEP   IT   LIQUID!

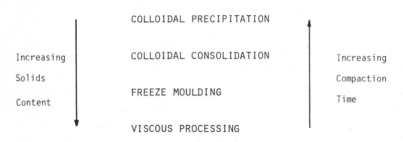

COLLOIDAL PRECIPITATION

Increasing   COLLOIDAL CONSOLIDATION   Increasing

Solids   Compaction

                             FREEZE MOULDING

Content   Time

                             VISCOUS PROCESSING

Fig. 5. Solution processing techniques for the production of technical ceramics.

compacts will result in the uncontrolled development of a porous microstructure with a wide grain size distribution.

Therefore, viable technologies for the green compaction of consistent high performance ceramics must be capable of avoiding or dispersing powder agglomerates and creating shape and form without subsequent reagglomeration. Solution techniques are being investigated to meet these requirements (Fig. 5). Under ideal conditions the powder should be synthesised in a liquid environment which it should not leave until the forming process is complete.

Colloidal chemistry has produced monosized spherical precipitates by the controlled hydrolysis of metal alkoxides.[5] These precipitates become monosize sub-micrometre powders, which can subsequently be consolidated into green compacts by gravitational settling from a dilute stable dispersion (usually in water). However, there are considerable problems arising from the reactivity of alkoxides, the very long sedimentation times and the elimination of defects in the assembly of colloidal powders throughout the volume of the green body. This technique may be adopted for the production of thin ceramic sheets, but it is unsuitable for complex shapes.

Slip casting techniques are generally able to disperse soft powder agglomerates and then form shapes by colloidal consolidation. Improvements to the basic casting process, including electrophoretic or pressure assistance, control the attraction between particles in suspension by modifying the surface energy of the particles or the hydrogen bonding capability of the liquid. Nonetheless, only relatively thin sections can be economically produced.

Freeze moulding[6] involves the injection of a medium solids content slip into a mould whose temperature is below the freezing temperature of the liquid. After demoulding, the liquid is removed by freeze drying. This technique allows the dispersion of soft powder agglomerates in relatively concentrated solutions, followed by pressure assisted particle assembly into complex forms using freeze moulding and drying to suppress the formation of liquid bridges or gaseous voids between the particles. At present, freeze moulding is not used for volume production of ceramics.

Viscous processing uses high shear mixing to break down agglomerates and coat individual particles with just sufficient quantities of processing aids to facilitate uniform compaction of coherent complex green shapes. For the injection moulding of ceramic powders, 'doughs' of powder plus organic binders have been prepared in twin screw polymer extruders.[7] Alternatively, ceramic bodies suitable for extrusion or sheet moulding have been produced using twin roll mills and water based polymer binder systems.[8] With agglomerated powders, both techniques give prefired densities in the green compact which are significantly greater than those obtainable by dry pressing. The principles of both techniques are now being used in the production of high grade technical ceramics.

5. KILNING

The sintering process involves the competing mechanisms of densification and grain growth within the ceramic microstructure. Conventional kilning technology is focussed on achieving the correct balance from a combination of a uniform green density and the controlled uniform increase in temperature. Additional refinements may include changing the furnace atmosphere (reducing, vacuum) and using grain-growth inhibitors (for example magnesium oxide in alumina).

For the established oxide ceramics, such technology is very flexible and cost effective. However, the newer high performance ceramics need more sophisticated kilning technology to achieve consistently their superior properties.

Magnesia partially stabilised zirconia requires heat treatment after densification to produce the phase changes necessary for its characteristic

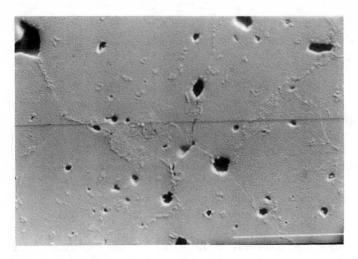

Fig. 6. Inter-granular porosity in magnesia partially stabilised zirconia ceramic with a fired density of 5·6 gcm⁻³.

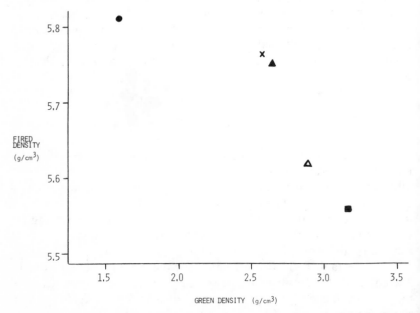

Fig. 7. Effect of powder type on densities of magnesia partially stabilised zirconia ceramic. ●, chemical vapour dissociation; ×, plasma dissociated mineral; ▲ (fine), △ (medium), ■ (coarse), graded powders from chemical-thermally decomposed mineral.

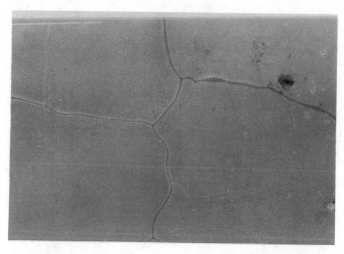

Fig. 8. High density microstructure (fired density $5.8\,\mathrm{gcm^{-3}}$) of magnesia partially stabilised zirconia produced from chemically synthesised sub-micrometre powder.

transformation toughening.[9] This involves increasing the temperature from 1400° C, where the ceramic is densified, to greater than 1700° C, that is into the cubic phase field. This is followed by controlled cooling to about 1000° C, which governs the nucleation of the tetragonal phase, and subsequent 'ageing' in the temperature range 1000–1400° C for growth of the tetragonal phase precipitates. This temperature regime is unlikely to be obtained in conventional large ceramic kilns; custom furnace technology is necessary. Furthermore, the significant increase above the densification temperature results in excessive grain growth (mean size 50 μm) and consequent entrapped porosity in the final sintered ceramic (Fig. 6). Therefore, the achievement of maximum density by the use of fine powders and non-agglomerate processing is very important. The influence of powder type on the green and sintered density of magnesia partially stabilised zirconia is shown in Fig. 7. The highest sintered densities are obtained from the chemically synthesised sub-micrometre powders (Fig. 8).

The nitride ceramics pose additional kilning problems. In particular, the densification of silicon nitride by self diffusion is extremely slow, and the potential benefits of using higher temperatures and liquid phase sintering are negated by dissociation of the nitride and the volatility of the oxide sintering aids at temperatures above 1800° C. In

B. G. Newland

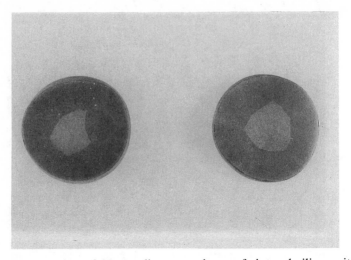

Fig. 9. Cross-section of 15 mm diameter spheres of sintered silicon nitride showing lower density core.

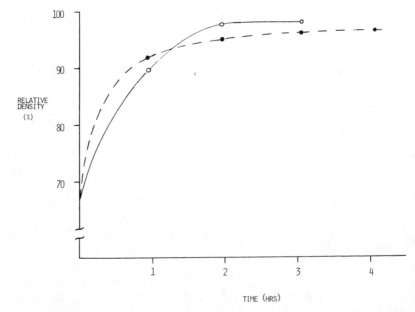

Fig. 10. Effect of nitrogen overpressure on the densification of inadequately sintered silicon nitride ceramic. ○ 1 MPa 1850 C; ● 5 MPa 1850 C.

the laboratory, these decomposition reactions are suppressed by sintering the green compacts in a bed of nitride powder which produces a local nitrogen overpressure. This remedy, however, does not produce full densification in thick sections (>10 mm), where the centre is often less dense than the outside (see Fig. 9). Increasing the nitrogen overpressure only accentuates the density differences (Fig. 10). It has been found necessary to first obtain uniform liquid phase sintering and then increase the temperature and nitrogen overpressure to achieve complete densification. This two-step 'gas-pressure' sintering process is used in the production of silicon nitride ceramic turbocharger rotors.[10]

6. NET SHAPING

The production of technical ceramics by green compaction and sintering of powders results in volume shrinkage, which makes the close control of as-sintered dimensional tolerance difficult to achieve despite the processing improvements described above. Net-shaping techniques have been developed which apply forming pressure at the sintering temperature (Fig. 11). They have the added benefits of producing virtually theoretical density in the finished ceramic.

Hot-pressing is an established net-shaping production process, but its use of uniaxial compaction and graphite tooling limits compaction shape and complexity. Now it is only competitive for the production of small batches and/or large components of monolithic ceramics. It remains, however, the most effective process for the full consolidation of ceramic composites which use materials with differential densification characteristics, for example alumina-silicon carbide whisker reinforced cutting tool tips.

Hot-isostatic-pressing (hipping) of technical ceramics uses high gas pressures and temperatures, for example 100–200 MPa, 2000°C for silicon nitride, to achieve net shaping by either sintering and densification of encapsulated green compacts, or densification of pre-sintered blanks. Despite the high cost of purchasing and maintaining equipment, hipping is potentially more cost effective than hot-pressing for the volume net-shape production of small, dense ceramic components, for example ball bearings and cutting tool tips.

However, although hot-pressing and hipping can produce high sintered densities with any defects remaining very small, they cannot remove processing impurities which could act as critical flaws.

A. HOT PRESSING

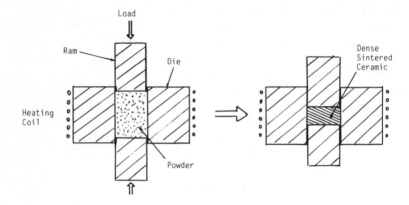

B. HOT ISOSTATIC PRESSING

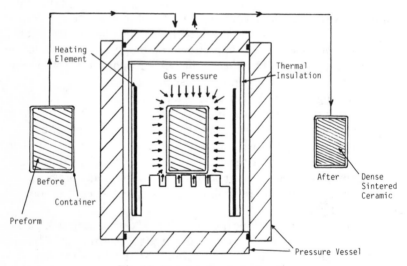

Fig. 11. Net shaping techniques.

7. MACHINING

Although careful control of processing conditions or the use of net shaping techniques can produce ceramic components with consistent dimensional accuracy, for many applications the tolerances required are so small that machining will be necessary after sintering.

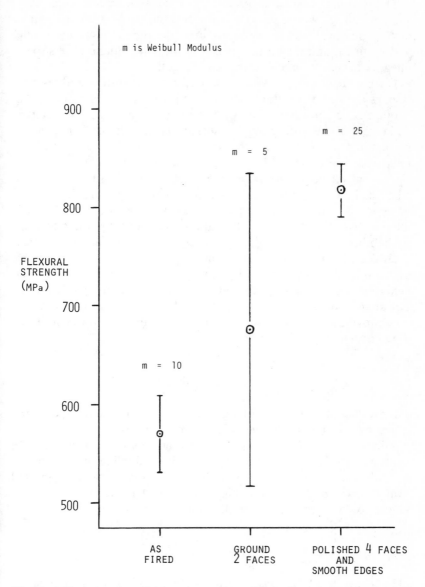

Fig. 12. Effect of surface finish on strength and Weibull modulus of a zirconia ceramic.

The shaping of sintered ceramics is usually considered in terms of 'machinability', which can be quantified in terms of, for example, material removal rate, power or grinding force required, surface profile, surface finish and tolerances. The effect of machining on the surface integrity of the ceramic is sometimes forgotten.

During machining, flaws are introduced on the surface and residual stresses are generated in the surface and subsurface. The machining damage zone can extend up to $100\,\mu$m below the surface. The nature and extent of the surface and subsurface damage varies from piece to piece. If the sintered ceramic has been produced with a dense uniform microstructure, as described above, then the surface damage will have a significant effect on the fracture strength; variable surface damage resulting in a low Weibull modulus. Surface damage can subsequently be removed by fine machining and polishing, except on sharp edges. If these are smoothed, however, then the surface contribution to fracture probability is minimised. Using this approach, high grade technical ceramics which have been machined to close tolerances can exhibit high Weibull moduli (see Fig. 12). To date, this has seldom been achieved due to lack of purpose-built ceramic machining equipment. Fortunately, the growing demand for high grade technical ceramics is now providing the economic justification for the development of low damage machining techniques.

8. CONCLUSIONS

The production technology of high grade technical ceramics is entering a period of major change. Significant improvements have been obtained in powder quality and processing consistency which can be combined to give ceramics of high integrity and reliability. At present the adoption of these improvements is restricted by their limited availability, small scale and high cost. The demands of the market place, however, are rapidly eroding these barriers.

REFERENCES

1. Davidge, R. W., *Mechanical behaviour of ceramics*, Cambridge, Cambridge University Press, 1979.
2. Reed, J. S., Carbone, T., Scott, C. and Lukasiewicz, S., *Processing of crystalline ceramics*, New York, Plenum Press, 1978, p. 171.

3. Van de Ven, T. G. M. and Hunter, R. J., *Rheologica Acta,* **16** (1977) 534.
4. Lange, F. F., *J. Am. Ceram. Soc.,* **67** (1984) 83.
5. Barringer, E. A. and Bowen, H. K., *J. Am. Ceram. Soc.,* **65** (1982) C-199.
6. Hausner, H., Proceedings of The Second International Symposium on Ceramic Materials and Components for Engines, Bad Honnef, Deutsche Keramische Gesellschaft e.V., 1986, p. 27.
7. Edirisinghe, M. J. and Evans, J. R. G., *Proc. Br. Ceram. Soc.,* **38** (1986) 67.
8. Kendall, K., McNalford, N. and Birchall, J. D., *Proc. Br. Ceram. Soc.,* **37** (1986) 255.
9. Pascoe, R. T., Hannink, R. H. J. and Garvie, R. C., *Science of ceramics, Vol. 9,* Faenza, Ceramurgica, 1977.
10. Hattori, Y., Tajima, Y., Yabuta, K., Matsuo, J., Kawamura, M. and Watanabe, T., Proceedings of The Second International Symposium on Ceramic Materials and Components for Engines, Bad Honnef, Deutsche Keramische Gesellschaft e.V., 1986, p. 165.

6

Nondestructive Evaluation of Advanced Ceramics

K. GOEBBELS

Tiede Rissprüfanlagen, GmbH D-7087 Essingen bei Aalen, FRG

ABSTRACT

Advanced ceramics have several advantages which make them important materials for structural applications: low density, high strength and hardness, low heat expansion and good heat conductivity combined with high thermal stability and oxidation resistivity. However, one disadvantage is that even at high temperatures advanced ceramics are brittle, which means that stress peaks at defects cannot be released by plastic flow and ductility. From the stress analytic and fracture mechanical point of view for design loads of some 100 MPa, defects with linear dimensions of 10–100 µm — at the surface still smaller ones — will cause failure. This makes a nondestructive quality control absolutely necessary. Because conventional NDE-methods are able to detect and to characterize defects with dimensions of only about two orders of magnitude higher, special research and development efforts were made during the last decade to develop NDE-techniques with high sensitivity and high resolution.

The full variety of NDT-techniques described in the different chapters has demonstrated their ability for quality control of advanced ceramics mostly in laboratory applications and in limited areas of application. Only for high resolution X-ray testing in projection technique does the state of the art of the equipment allow a direct application in practice. However, here the film is the up-to-now accepted receiver medium. The filmless technique, using X-ray sensitive cameras, image intensifiers and other types of solid state detectors, still needs further optimization. For ultrasonic testing, as well as for photoacoustic and vibration analysis, the appropriate equipment is still under development. However the realization of apparatus and sensors that are applicable in practice under economical considerations can be foreseen for the next few years.

1. INTRODUCTION

1.1 Advanced Ceramics: Materials and Components

What is the meaning of 'Advanced Ceramics'? During the last 15 years worldwide an enormous effort was made to develop ceramic materials, to design and to fabricate ceramic components for special applications. Advanced ceramics, fine ceramics, structural ceramics, high-tech ceramics and other terms are used to describe ceramics based on extremely pure, composition-controlled and ultra-fine grained particles of aluminium oxide, silicon nitride, silicon carbide, zirconium dioxide and others. They are used as electro-ceramics, e.g. for integrated-circuit substrates and packages, and as structural ceramics, e.g. for cutting tools, catalyst carriers, engine and turbine parts, for high-temperature, corrosion and wear resistant machinery parts. Table 1 reviews important ceramic materials and components in this field.

1.2 Demands to Nondestructive Evaluation: Defect, Structure, Stress

One common aspect of these parts is their failure as a result of inadequate quality. The brittleness of ceramics — due to the largely covalent nature of the bonding forces — is responsible for the fact that stresses due to mechanical or thermal loading cannot be released by plastic flow. Peak stresses at defects especially are therefore failure-initiating effects. In this context the single grain ($\lesssim 1\,\mu$m) of a poly-crystalline material as well as a foreign inclusion or a pore of this size have to be considered as the *smallest* defects and fracture-mechanics calculations will be used to define *critical* defect sizes as shown in Fig. 1 for hot-pressed silicon nitride (HPSN) by Evans.[1] The main point of this figure is that defects down to $100\,\mu$m size will cause failure for typical loads of 300 MPa depending upon the type of the defect. Experiments done with reaction bonded silicon nitride (RBSN) showed that in four-point bending tests surface defects of $\geqslant 15\,\mu$m were causing failure at loads above 200 MPa.[2] So nondestructive testing (NDT) methods have to detect and to quantify defects with dimensions down to some micrometres. This demand is at least two orders of magnitude below the present state-of-the-art in NDT.

TABLE 1
Advanced ceramic materials and components

Field of Application	Component	Material
Structural ceramics	Cutting and drawing tools	Al_2O_3, ZrO_2, Si_3N_4, SiAlON + Y_2O_3), PSZ (partially stabilized zirconia (ZrO_2))
	Seal	SiSiC, CSiC
	Valve	Al_2O_3, Si_3N_4
	Bearing	PSZ, Si_3N_4
	Catalyst carrier	MAS (magnesium aluminium silicate)
	Burner	SiSiC, Si_3N_4
	Sodium vapour lamp	Translucent Al_2O_3
	Coating	ZrO_2, TiN, WC
Bioceramics	Hip endoprostheses	Al_2O_3, CSiC
	Dental implant	Al_2O_3
Electroceramics	IC-package	Si_3N_4
	Sensor	ZrO_2, TiO_2, SnO_2
	Piecoelectrics	$BaTiO_3$, PZT (lead zirconate titanate)
Energy conversion	Heat exchanger tube	Al_2O_3, SiC, SiSiC
	Na-S-Battery tube, collar	β-Al_2O_3, α-Al_2O_3
Engines	Pre-combustion chamber	ATI (Al_2TiO_5), PSZ, Si_3N_4
	Turbocharger rotor	SiC, Si_3N_4
	Piston head, cylinder liner	ZrO_2, PSZ, ATI
	Portliner	ATI
	Valve seat, camshaft, bearing, gliding element	ZrO_2
Gas turbines	Combustion chamber, inlet cone, stator	SiSiC, Si_3N_4
	Rotor	SiC, Si_3N_4
	Heat exchanger	SiSiC, Si_3N_4, MAS, LAS (lithium aluminium silicate), AS (aluminium silicate)

Two additional features describing a material's ability to withstand the loads they are used and fabricated for, are microstructure and residual stress.

The strength of ceramics strongly depends upon the material's

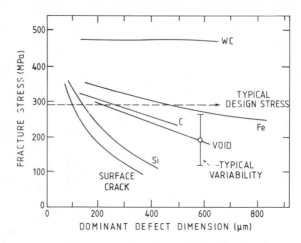

Fig. 1. Dependence of fracture strength on defect size for fracture initiating
defects in hot pressed silicon nitride (after Evans[1]).

microstructure, including the grain size (distribution), the porosity (pore
size (distribution) and percentage of porosity), the homogeneity
(described integrally by the density ρ) as well as the fabrication induced
anisotropy. Even for 'green' ceramics and semi-finished products (e.g.
plates) a check referring to these parameters can save time and money if
the quality is not appropriate.

The local stress situation in a component is given by the sum of
residual and applied stresses. The determination of residual stresses at
the surface as well as for the bulk is therefore also an important demand.
Surface finishing, for example, can result in residual stresses reducing
the strength by up to 40% as reported by Marshall.[3]

Table 2 gives some of the important strength controlling effects in
ceramics which have to be covered by NDT and the interpretation of
NDT results (i.e. nondestructive evaluation, NDE).

The important role of quality control by NDT for ceramics can be
emphasized by comparison with the basic safety concept for nuclear
power plant components:

— The mean grain size of a reactor pressure vessel steel is about
   10 μm. The size of defects which have to be detected is 3 mm, i.e.
   about two orders of magnitude higher. The defect size which will
   cause failure is roughly 100 mm referring to fracture mechanics
   calculations, which is again more than one order of magnitude
   above this value.

TABLE 2
Strength controlling effects in ceramics

Parameter	Parameter changes during		
	Production	Machining	In-service
Structure			
1. Density	Density variations of green ceramics, porosity		
2. Homogeneity	Inhomogeneous green bodies, porosity variations, coarse grained zones	Inhomogeneous bonding layers	Corrosion/oxidation
3. Isotropy	Anisotropic green bodies, anisotropy due to manufacturing		
Stress			
1. Residual stresses	Generally due to manufacturing, locally at heterogeneities	Surface machining, bonding	
2. Load stresses			Temperature gradients, bonds
Defect			
1. Surface	Cracks, pores, inclusions	Cracks, corner chips	Cracks, crack growing, corrosion of bonding layers, erosion
2. Volume	Cracks, pores, inclusions, shrinkage cavities	Cracks, delaminations (bonds)	Cracks, reaction of inclusions with surrounding matrix material

— The mean grain size of engine ceramics is about 1 μm and the size of defects which will be responsible for failure is about 100 μm. The detection limit for defects has to be sufficiently lower than this.

This comparison gives an impression of the narrow safety margin referring to ceramic parts. Without detailed knowledge about a defect detected, only little value will be obtained from NDT results. But with additional information about the kind of defects, their size and their orientation this margin can be improved and less material has to be rejected.

2. PHYSICAL BACKGROUND OF ULTRASOUND AND X-RAY RADIOGRAPHY

The interaction between a defect and the analysing quantity (mechanical waves, electromagnetic fields, high energy radiation, etc.) can be described by a few parameters. These are mainly:

— the difference in physical properties of defect and surrounding material; and
— the relations defect dimension to analysing wavelength d/λ, defect dimension to analysing beam diameter d/D and defect dimension to 'imaging' geometry parameters.

To obtain maximum values for both parameter groups results is the goal of optimized detection and evaluation. Ultrasonic waves and X-rays have up to now the highest potential for NDE of ceramics. Therefore, it seems to be necessary to describe their physical background before discussing their applications to special problems and components.

2.1 Ultrasound
The advantages of ultrasonic methods are

— the ability to adjust the wavelength λ by the right choice of the frequency f for the optimal interaction with defects, microstructures and stresses ($\lambda = v/f$, v = velocity).
— the possibility of using free waves (L = longitudinal waves, T = shear waves) as well as guided waves (R = surface waves;

TABLE 3
Ultrasound velocities of advanced ceramics and of some typical defects

Material	$\rho(\text{Mg/m}^3)$	$v_L(\text{m/ms})$	$v_T(\text{m/ms})$
Silicon nitride			
RBSN	2·7	9·0	5·2
SSN	3·1	10·5	5·9
HPSN	3·2	11·0	6·1
Silicon carbide			
RBSiC	3·0	9·5	6·5
SSiC	3·1	12·0	7·5
HPSiC	3·2	12·1	7·7
SiSiC	3·1	11·4	7·6
Others			
Al_2O_3	4·0	10·6	6·3
ZrO_3	5·8	7·0	3·7
Defects			
SiO_2	2·6	6·0	4·1
Fe	7·8	5·9	3·2
Si	2·3	9·0	5·3
WC	15·0	6·7	4·0
C	2·3	3·7	2·0

$$v_L = \sqrt{\frac{E(1-v)}{\rho(1+v)(1-2v)}}, \qquad v_T = \sqrt{\frac{G}{\rho}}, \qquad v = \frac{2-(v_L/v_T)^2}{2-2(v_L/v_T)^2},$$

$$v_R = \frac{0·87 + 1·12}{1+v} \cdot v_T \approx 0·9\, v_T$$

additionally plate waves, tube waves, bar waves) for the appropriate problem solution.

Table 3 gives characteristic mean values of $v(L,T,R)$ for different advanced ceramic materials.

The interactions of ultrasonic waves and pulses with matter are twofold: elastic and anelastic. The elastic interaction holds the character of mechanical vibrations and is given by wave propagation with the velocity v in an undisturbed medium by reflection, diffraction and scattering at locations where the 'sound impedance' (ρv) changes. The anelastic interaction describes the dissipation of wave energy during propagation by absorption. The attenuation coefficient α of a wave is the sum of absorption (α_A) and scattering (α_S).

2.1.1 Velocity

The relationship between $v(L,T,R)$ and the elastic moduli E,G,v (elastic modulus, shear modulus and Poisson's ratio, respectively) as well as the density ρ is well known and given in Table 3. Important facts are:

— v is a function of modulus and density, so velocity changes can be caused by several and differently working microstructural effects;
— if modulus and density are changing in the same direction (increasing, decreasing), the velocity can remain unchanged;
— second-order effects upon the velocity, such as stresses, texture and scattering, are physically understood and can be used for material characterization.[4]

The elastic anisotropy of single crystals, that is, direction dependent velocity, will cause macroscopic anisotropy if the crystal orientation is not random. Because of its relation to the velocity, anisotropy can be determined by velocity measurements with wave propagation of free and/or guided waves in different directions.[4]

Additionally, the ultrasonic wave velocity is stress-dependent and therefore can be used especially to determine residual stresses. The most appropriate technique seems to be the birefringence measurement using polarized shear waves.[5,6] To separate the effects of texture and stress for the evaluation of both, the birefringence dispersion, that is frequency dependence of the double refraction effect, provides the necessary additional measurement parameters.[6]

2.1.2 Attenuation

The propagating ultrasonic wave loses its energy by absorption and scattering. During absorption several interactions are able to change the vibration energy finally into heat: thermoelastic losses, dislocation damping, viscoelastic losses and absorption phenomena on an atomic and subatomic scale.[7] So in technical materials the absorption coefficient α_A gives only a qualitative indication of the material's absorption behaviour. Above all α_A is proportional to the frequency f.[8]

Scattering of ultrasound occurs if the propagating wave feels changes in the sound impedance $Z = \rho v$. The vibrational character of the energy

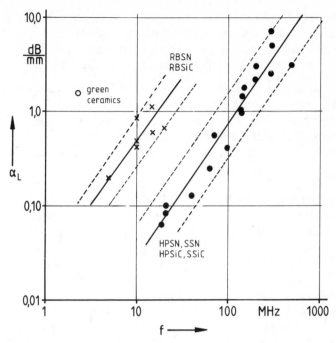

Fig. 2. Attenuation of longitudinal waves in different types of advanced ceramics.

is maintained but the scattered energy is going into the whole space angle, therefore an energy loss results for the propagation direction. The amount of scattering described by the scattering cross-section γ of a single scatterer and/or the scattering coefficient α_S (where $\alpha_S = n_o\gamma/2$ for isolated scatterers and neglecting multiple scattering), is influenced by

— the change ΔZ at grain boundaries, phase boundaries,
— the relationship linear scatterer dimension d to wavelength λ; and
— volume concentration n_o of the scatterers.[8]

Figure 2 gives typical values of attenuation as function of frequency in advanced ceramics.

2.1.3 Sensitivity and resolution

Any interaction between ultrasound and matter (microstructure, defects) follows the well known behaviour sketched in Fig. 3:

— small effects causing small signal amplitudes for $d \ll \lambda$, but strongly increasing with d/λ (Rayleigh region);

— end of monotonically increasing interaction for $d \approx \lambda$;

— system resonances and geometrical reflection for $d \gg \lambda$.

Highest sensitivity therefore will be reached in the Rayleigh region while the highest signal amplitudes are measured for $d \gtrsim \lambda$.

Referring to velocity measurements, the resolution is given by the time-of-flight measurement method, today mostly phase-sensitive pulse superposition measurements are applied,[9] allowing accuracies of 1×10^{-4} and better.

According to the spatial resolution, the sound beam cross-section determines the lateral resolution while the pulse length determines the axial resolution. Both in principle can reach the wavelength using axially either synthetic aperture or short pulses with broad frequency spectra, and using laterally focussed beams with curved transducers, phased arrays, acoustic holography or synthetic apertures.[10]

2.1.4 Defect detection and characterization

As shown in Fig. 3, maximum signal amplitudes are obtained for defect sizes comparable to or greater than the wavelength λ. The high value of v in advanced dense ceramics makes frequencies of 100 MHz and above necessary for the characterization of defect sizes around $100\,\mu m$ and smaller. Therefore, high-frequency ultrasonics were

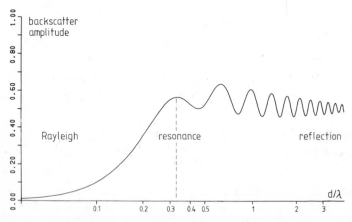

Fig. 3. Interaction between ultrasound and scatterers of linear dimension d (calculated for a pore in hot pressed silicon nitride).

developed with upper values of about 1 GHz, resulting in wavelengths of 10 μm and smaller. Because of attenuation α increasing with f (linearly for α_A and with some power of f for α_S) high-frequency ultrasound pulses are propagating only some wavelengths below the surface. In any case of application it is a question of optimization between sensitivity/resolution on the one hand and the attenuation/ penetration depth on the other. Developments obtained for special applications are discussed in the subsequent chapters where they are relevant.

One exception in the need to reach $d \approx \lambda$ can be made for guided waves, e.g. surface waves or plate waves. The character of a guided wave is very sensitive to disturbances. Therefore, in this case an appropriate defect detection is already given for $d/\lambda \gtrsim 0.01$.[11] This immediately allows the use of 'low' frequency ultrasonic waves, e.g. 20 MHz, to detect defects of sizes just below 10 μm with advantages concerning equipment, coupling between probe and specimen and attenuation.

2.2 X-ray Radiography

Single defects as well as extended structural heterogeneities (density variations, anisotropy of the microstructure, stresses, phase boundaries and bonding areas) can be imaged and analysed with X-rays. Referring to single defects, detectability (contrast), visualization and spatial resolution are described by a few equations.

To detect a defect it is necessary to have a certain image contrast between the flaw and the surrounding medium. This contrast K is given by:[12]

$$K = \text{const.}\Delta\mu.d_z/(1 + I_S/I_D)$$

where the constant includes receiver parameters (film, image intensifier, etc.), d_z = defect size in X-ray beam direction, $\Delta\mu$ = difference in X-ray absorption for matrix and defect, I_S = scatter intensity, I_D = image intensity. Some relevant $\Delta\mu$-combinations for advanced ceramics can be derived from Table 4.

The 'density resolution equation' — density ρ proportional to absorption μ for low energy X-rays — is given by[13]

$$\frac{\Delta\mu}{\mu} = \frac{1}{\log_e} \cdot \frac{K}{\gamma} \cdot \frac{1}{\mu d_z} (1 + I_S/I_D)$$

where γ = film gradient. Relative density changes $\lesssim 0.01$ can be resolved. The 'thickness resolution equation' can be seen as the analogue to the density resolution equation:

TABLE 4
X-ray absorption coefficients
(maximal energy 50 kV)

Material	$\mu(\text{cm}^{-1})$
HPSN	2·56
HPSiC	2·87
Si	2·92
Fe	58
C	0·51
$AgNO_3$	89

$$\frac{\Delta d_z}{d_z} = \frac{1}{\log_e} \cdot \frac{K}{\gamma} \cdot \frac{1}{\mu d_z} (1 + I_S/I_D)$$

Again relative thickness changes $\gtrsim 0.01$ can be resolved.

To visualize defects for the naked human eye their image must be roughly 100 μm in diameter with the contrast given above. Especially for defects in ceramics, where a defect size significantly smaller than 100 μm has to be found, magnification is unavoidable. With respect to industrial applications secondary enlargements are time consuming and expensive. Therefore, the so-called projection technique, Fig. 4, is

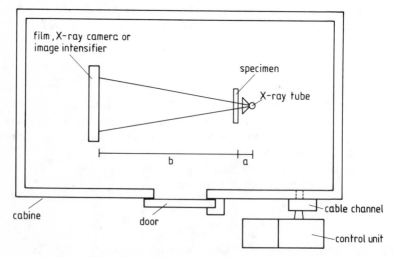

Fig. 4. Microfocus X-ray unit with projection technique.

the state-of-the-art method with direct or primary magnifications m up to $300\times$ and more:[14]

$$m = \frac{a + b}{a}$$

$$D_1 = md_1$$

where a = distance X-ray source to defect, b = distance defect to receiving medium, d_1 = lateral defect size and D_1 = lateral defect size in the image plane. In any case, in conventional contact radiography or in projection technique, $m \geqslant 1$.

But the sharpness with which a defect will be imaged on a film or a screen is defined by the 'geometrical unsharpness' U_g, which is proportional to the focal point size s of the X-ray tube, Fig. 5:

$$U_g = s(m - 1)$$

U_g results in a shadowing area along the contour of a defect. A clear defect image needs $U_g < D_1$. While in conventional contact radiography m is close to 1 and therefore U_g close to zero, this is inappropriate for the visualization of defects smaller than $100 \, \mu$m where $m \gg 1$ is a necessary boundary condition.

But a large m means $m \approx (m - 1)$, so

$$U_g/D_1 \approx s/d_1$$

A small U_g value therefore can only be obtained by a small s value. During the last 5 years several manufacturers of X-ray equipment have developed X-ray sources with spot sizes down to $s \approx 1 \, \mu$m. Microfocal X-ray instrumentation now belongs to the state-of-the-art.

In the image plane of a film two neighbouring defects can be resolved laterally if their contours are separated by about $2U_g$:

$$\frac{\Delta d_1}{d_1} \geqslant 2 \frac{U_g}{d_1}$$

For filmless radiography, e.g. computerized tomography (CT), detectability $\Delta\mu/\mu$ and lateral resolution are related to each other:[13]

$$\left(\frac{\Delta\mu}{\mu}\right)^2 \cdot \left(\frac{\Delta d_1}{d_1}\right) = \frac{\pi^2}{4} \cdot \left(\frac{1}{\mu x}\right)^2 \cdot \frac{1}{3N_{total}}$$

where x is the pixel size and N_{total} = dose ($\approx 10^6$).

For industrial applications, in general, X-ray CT is a solution which

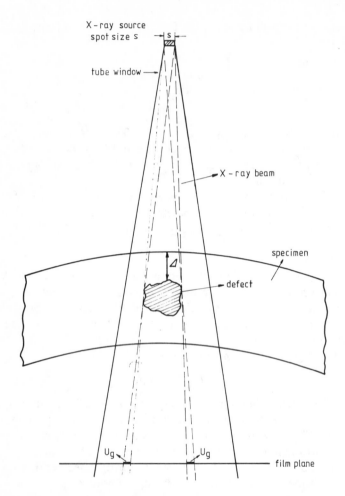

Fig. 5. Definition of geometrical unsharpness U_g in X-ray radiography.

is too expensive and up to now too time consuming. The film on the other side is expensive, too, regarding complex shaped ceramic parts which need many images obtained from different directions and optimized for different thicknesses through-radiated. Therefore, the visualization of structure and defects by filmless techniques using X-ray sensitive cameras or image intensifiers has been improved. While this technique alone is not as sensitive as film and CT, signal processing of digitized data (averaging over 2^n images, post processing by zero-image correction, filtering and other methods) allows these values to be nearly reached.[15]

Texture and stress measurements by X-ray methods belong to the well known state-of-the-art and need no separate description. This can be taken from the literature as well as the principle and the limits of X-ray CT. For technical parts so-called transaxial tomography is appropriate, rotating the object to be analysed and not the X-ray source and the detectors.[13]

3. GENERAL MATERIALS CHARACTERIZATION

3.1 Ultrasonic Testing

Semi-finished parts such as plates, discs or rings are tested, with reference to their homogeneity, easily by ultrasonic velocity measurements. Variations of v along and across a component are related to density variations and elasticity variations and therefore the velocity is a sensitive indicator of local structural changes or of changes due to the fabrication procedure.[16] Figure 6 gives an example for a velocity

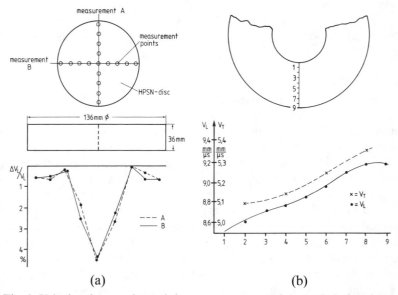

(a) (b)

Fig. 6. Velocity changes due to inhomogeneous materials. (a) Relative change of longitudinal wave velocity across a disc of hot pressed silicon nitride (HPSN). (b) Change of longitudinal and shear wave velocity across a disc of reaction bonded silicon nitride (RBSN).

variation due to an incomplete densification of HPSN and for a velocity gradient due to a nitridation process gradient in RBSN. The advantages of such tests are:

— use of 'low' frequency ultrasound ($\ll 100$ MHz);
— automation of procedure in immersion test beds,
— relatively fast testing procedure,
— materials characterization in an early stage, to avoid further expensive finishing, if the quality is insufficient.

Another qualification with ultrasonic waves of frequencies $\ll 100$ MHz is the anisotropy measurement, where the anisotropy can be due to microstructural elastic anisotropy (texture) and/or due to stress anisotropy (residual stress). Both effects lead to direction dependent velocities, especially of linearly polarized shear waves. As for the homogeneity characterization, the result at a given point of measurement is an integral over the sound path with a lateral resolution of the sound beam diameter.

While texture can lead to relative differences in sound velocities of a few percent, residual stresses show effects smaller by one order of magnitude. Because the third-order elastic constants of polycrystalline silicon carbide, silicon nitride, aluminium oxide, zirconia, etc., are not known up to now, only a qualitative mapping of 'stress anisotropy' is possible.

3.2 Vibration Analysis

The excitation of mechanical vibrations with long wavelengths — where the wavelength corresponds roughly to the dimensions of the object under test — leads to resonance vibrations which can be measured by transducers, or simply microphones. The simplest solution uses an impact excitation — with a broad frequency spectrum — and analyses the specimen-related main vibration mode (resonance frequency). For simple geometries (ball, bar, disc) these values are related to elastic modulus, density and geometry; for more complex shaped components such as turbine blades or whole rotors they are parameters characterizing integrally the part. Figure 7 shows the main resonance frequencies of three series of RBSN blades as a function of the weight and of two series of RBSN four-point-bending-test specimens as a function of the ultrasonic wave velocity v_L. Different types of correlation tendencies as well as different scatter bands are clearly seen.[17]

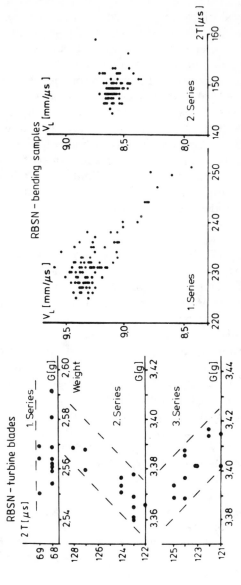

Fig. 7. Vibration analysis of different series of RBSN specimens compared to weight G and longitudinal wave velocity v_L (resonance frequency = $1/T$).

For the mass production of equally shaped parts this seems to be an appropriate tool for the rapid and integral analysis of each single component. The method seems to be a sensitive indicator, too, for defects occurring during service: silicon nitride turbine blades with resonance frequencies of about 30 kHz for the as-manufactured state show a significant decrease (18 kHz), if cracks are present after oxidation treatment.[15] Even more information can be obtained by vibration analysis, if not only the main resonance frequency but the whole resonance spectrum is evaluated ('modal analysis'). The excitation again could be by an impact, but more resolution will be gathered by continuous wave excitation and frequency wobbling.[18] Position, amplitude and shape of ground modes and higher harmonic vibrations are related to locally occurring structure, stress and defect situations in the object. But the potential as well as the limits of vibration and modal analysis are not yet quantified. This belongs to outstanding extended nondestructive and destructive studies.

3.3 Radiography

There is no question that X-ray radiography also shows immediately on films or screens significant variations in the macroscopic structure of any object.[15, 17] Especially density variations are easily imaged because of the proportionality between ρ and μ. The advantage of X-ray CT is the possibility of quantitative density mapping with a resolution of about 0·1%. Volumetric defects such as pores or inclusions in general are also insensitive to the object orientation relative to the direction of radiation: their projection onto the film or screen is uncritical. Cracklike defects on the other hand are very sensitive to their orientation relative to the radiation: only oriented parallel to the beam, they will give an appropriate image with a contrast following the equations given under Section 2.2. Even a misalignment of about 15° and more will make them invisible, a situation which makes it necessary to take radiographs from many different orientations or to apply X-ray CT or to concentrate on directions where cracks mainly occur.

3.4 Proof Testing

Because NDT for advanced ceramics is still under development, today proof testing belongs alongside a simple visual inspection to the mostly used NDT methods. But there are several arguments against proof tests, even if they are narrowing scatter bands of parameters such as described in the literature:[19]

— The proof load does not correspond to the complex situation in practice: locally varying multiaxial stress states as well as temperature distributions cannot be simulated by the proof test.
— The result of a proof test will be that any remaining defect has a size smaller than a defect which will cause failure for the applied proof load. But there is no information how many smaller defects are present and how near they are to the size of a critical defect.
— Finally, there is no information if defects were generated during the proof loading or if others were growing during the test. In practice such defects especially could show further growth up to failure of the whole part in service.

So proof testing provides only a negative statement (no defect bigger than . . .) but no information about the real situation in the object and therefore the proof test remains a weak link in the chain of quality control which should be improved by real NDT methods such as those described in the following sections.

4. SURFACE AND NEAR-SURFACE TESTING

4.1 Visual Inspection by Liquid Penetrants

The highest loads generally occur at the surface, for mechanical forces as well as for thermal loads. Therefore, a surface defect from the fracture-mechanical point of view is the most severe flaw (see Fig. 1). Besides careful visual inspection, e.g. with microscope and endoscope, the liquid penetrant inspection belongs to the state-of-the-art (fluorescent liquids and conventional white/red liquid combinations). Problems arise during testing of reaction bonded materials because of the porosity (high noise background). Up to now, penetrant testing is not optimized for ceramics, so some uncertainties are given referring to the capillary effect in comparison to metallic materials for which the penetrants were developed. In principle, liquid penetrants can only visualize surface-breaking cracks, quantifying the crack length. Nothing can be found about the crack depth.

4.2 Ultrasonic (Surface) Waves and Acoustic Microscopy

Ultrasonic testing of surfaces has to be discussed under two aspects:

— testing of the surface, where the probe is coupled, with surface waves and with acoustic microscopy;
— testing of the opposite surface with shear waves.

Besides surface connected and near-surface inclusions only surface opening cracks are considered, because crack closure and cracks with corrosion products in the crack ground belong to up to now unsolved problems of quantitative NDE.

4.2.1 Surface waves

Surface or Rayleigh waves have the advantages that their interaction depth is roughly one wavelength λ_R and that by proper choice of the frequency f this depth can be varied widely:

$$\lambda_R = v_R/f$$

Examples of v_R of advanced ceramics can be derived from Table 3. But the most important fact referring to Rayleigh waves is their high sensitivity to defects as described under Section 2.1. This allows the use of low frequencies with low attenuation to detect flaws which are much smaller than the wavelength. Examples for near-surface defect and surface flaw detection are given in Fig. 8. The advantage of lower frequencies with lower attenuation and longer wavelengths is correlated with greater propagation lengths and less influenced by surface roughness.[20-23]

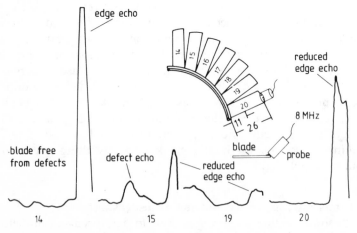

Fig. 8. Detection of surface defects (cracks) in RBSN rotor blades with ultrasonic surface waves (8 MHz, $\lambda_R \approx 0.5$ mm).

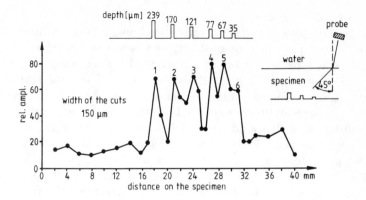

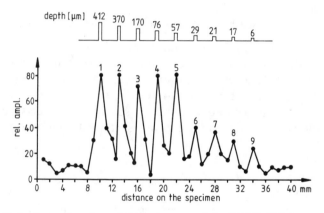

Fig. 9. Detection of surface defects (saw cuts) in RBSN (thickness 4·5 mm; 17 MHz, shear waves, 45°, $\lambda_T \approx 0·3$ mm).

4.2.2 Free waves

While the optimal reflection from defects occurs for $d \approx \lambda$, the ultrasonic testing for surface defects from the opposite side is relatively effective too with shear waves of oblique incidence, e.g. 45° and $d/\lambda \lesssim 1$. For $d > \lambda$ this comes from the so-called angle-mirror effect with nearly total reflection of the incident sound intensity. For $d < \lambda$ diffraction echoes ('crack-tip echo') of much lower intensity but with sufficient signal-to-noise ratio (SNR) are obtained especially from sharp defect edges.[23] Examples are given in Fig. 9. Focussing probes using lenses or curved transducers allow enhancement of the SNR. A focal diameter of about one wavelength is possible.

4.2.3 Acoustic microscopy

So far the results of ultrasonic testing are shown by A-scans (amplitude versus sound path or time-of-flight) or C-scans (mapping of defect positions in a plane parallel to the scanned surface when the signal amplitude overcomes a given threshold value). The need finally to approach $\lambda \approx d$ for the optimal defect detection and characterization leads to frequencies up to about 1 GHz and therefore wavelengths down to about $1\,\mu$m with sound beam diameters of about $10\,\mu$m. Careful scanning of surfaces with small beams makes the A-scan representation inadequate. Ultrasonic imaging, much used at frequencies around 5 MHz in medical diagnostics and metallic materials testing, becomes relevant. Because of the high resolution obtained and the necessary magnification, mostly on a TV screen, UT here is called acoustic microscopy.

Two principles (see Fig. 10) now belong to the state-of-the-art:[24, 25]

— SLAM (Scanning Laser Acoustic Microscope)
  The object is insonified by continuous wave (cw) ultrasound of 30, 100 or 500 MHz, while the resulting surface displacements on the opposite side are picked up with a scanned laser beam. In principle, the technique is mostly sensitive to defects near the surface along which the laser beam is scanned. Some examples are

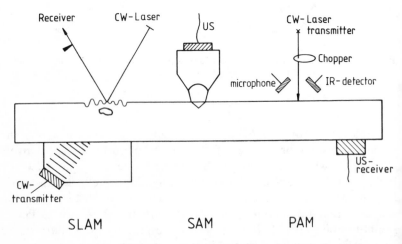

Fig. 10. Principles of Scanning Laser Acoustic Microscope (SLAM), Scanning Acoustic Microscope (SAM) and Photo-Acoustic Microscope (PAM).

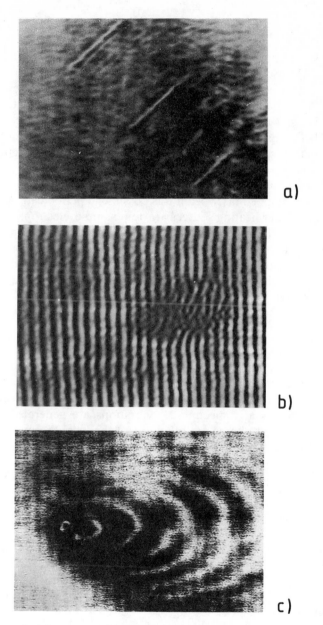

Fig. 11. Acoustic microscopy with the SLAM. (a) Saw cuts in RBSN (Fig. 9), 30 MHz; depth of the smallest cut is 17 μm. (b) Pore in sintered silicon nitride (SSN) taken in the interference mode at 100 MHz; the diameter of the pore is 500 μm. (c) Surface pore in HPSiC, 100 MHz, diameter of the pore is 100 μm.

shown in Fig. 11. But the method is also imaging volume defects.[26]
— SAM (Scanning Acoustic Microscope)
The surface is scanned with probes from 100 MHz to several GHz.
While the coupling comes from a small liquid droplet, the signal
imaged is gated from the focal point in the surface or slightly
below. Especially with the highest frequencies, acoustic microscopy
of this type is restricted to the surface or some wavelengths below.

4.3 Photoacoustic/Photothermal Microscopy (PAM/PTM)

With a chopped laser beam it is possible to periodically heat object
surfaces. Other heating sources are, for example, electron beams or X-
ray beams. The chopper frequency f lies roughly between 1 Hz and
1 MHz. The focussed beam allows a lateral resolution of some
micrometres during scanning a surface, while the analysed depth is a
function of the chopper frequency and is called thermal diffusion
length δ_{th} (Ref. 27)

$$\delta_{th} = \sqrt{\frac{k}{\pi \rho c f}}$$

where k = thermal conductivity, ρ = density, c = specific heat.

Table 5 gives some material constants for the calculation of
penetration depth values in ceramic materials. At the heated point
thermal waves and mechanical vibrations are generated and their
amplitudes and phases are functions of the local thermal properties.

TABLE 5

Materials constants for the calculation of thermal diffusion length and
impedance in photothermal microscopy

Material	$k(\mathrm{W\,m^{-1}\,K^{-1}})$	$\rho(\mathrm{Mg\,m^{-3}})$	$c(\mathrm{J\,g^{-1}\,K^{-1}})$
RBSN	18	2·7	0·85
SSN	22	3·1	0·85
HPSN	25	3·2	0·85
SSiC	75	3·1	0·9
HPSiC	90	3·2	1·0
SiSiC	100	3·1	1·0
Al_2O_3	33	4·0	0·8
ZrO_2	2·5	5·8	0·4
Al_2TiO_5	2·0	3·2	1·0

k = thermal conductivity; ρ = density; c = specific heat.

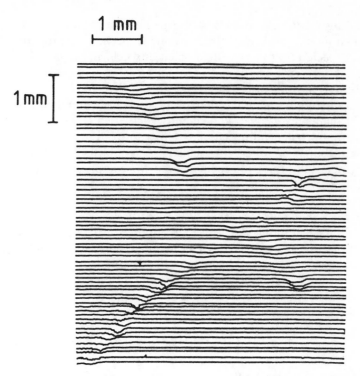

Fig. 12. Photoacoustic microscope image of surface breaking cracks in a SSiC turbocharger (phase-image, 33 KHz; 10 mm vertical displacement corresponding to 90° phase shift).

Therefore, on the receiving side infrared detectors, acoustic microphones, piezoelectric probes as well as laser interferometers can be used.[28] The method as a high sensitive surface analysing NDT technique is still under development but with special reference to ceramics. Results are known, for example, from ball bearings,[29] turbine blades and stator vanes,[30] ceramic coatings[31] and other types of specimens.[32] The principle is sketched in Fig. 10 while an example is shown in Fig. 12. It is to be expected that PAM/PTM will contribute significantly to ceramic components surface testing in the future.

4.4 X-rays

4.4.1 Texture analysis

The determination of texture by X-rays is well known and widely used in practice for metallic materials. To date there are no results regarding advanced ceramics in the open literature but, because of the severe influence of elastic anisotropy on the mechanical properties, it is to be expected that work will go on in this field in the near future.

4.4.2 Stress measurement

A comparable situation is that for the stress measurement, especially for residual stresses. Well known and widely used is the X-ray surface stress analysis of metallic materials. Up to now, only one report of X-ray stress measurements on ceramic materials was published,[33] demonstrating the general ability of the method even for complex shaped ceramic parts.

4.4.3 Dye enhanced radiography of surface defects

A simple technique can be used to improve the detectability and the sizing of surface opening flaws. Due to the capillary attraction for liquid penetrants (Section 4.1) cracks can be filled with a contrast medium, e.g. silver nitrate.[15, 34] The image then, as in Fig. 13 for example, shows the crack and its shape with much more contrast than without the agent.

5. BULK MATERIAL TESTING

5.1 High Resolution Radiography

5.1.1 Microfocus X-rays

For internal defects in complex shaped ceramic parts high resolution radiography with microfocus tubes and with projection techniques have the advantage over any other NDT method. Referring to the physics mentioned in Section 2.1, focal spot sizes s of down to $\approx 1\,\mu$m and direct magnifications m up to $\approx 300\times$ present structural features and defect details with enough sensitivity and contrast. Equipment with these characteristics belongs now to the state-of-the-art.

While the film up to now still seems to be the most sensitive recording medium, filmless radiography with image intensifiers, solid state cameras, X-ray sensitive photodiodes and others[35] will reach the film quality if appropriate digital signal processing is added.[15] This has to include, after the analogue-to-digital conversion (ADC) averaging over 2^n images ($n = 2,3,\ldots$), zero-image correction, contrast enhancing filtering as well as pseudo-colour display.

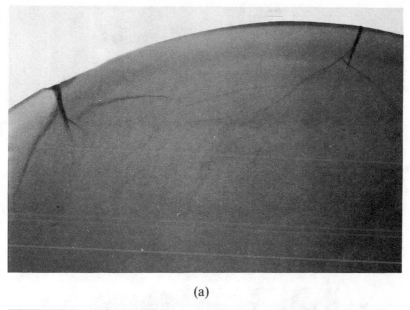

(a)

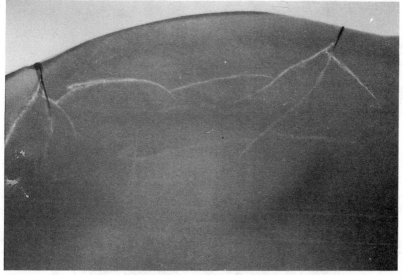

(b)

Fig. 13. (a) Microradiography of HPSiC-disc with cracks (film). (b) Same as before but contrast enhanced by $AgNO_3$.

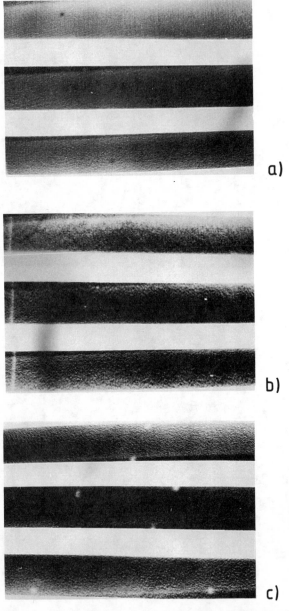

Fig. 14. Filmless received and image processed microradiographs of defects in RBSN; processing: image intensifier-ADC-averaging-zero image correction-contrast enhancement (filtering). (a) Pores 500 μm ϕ. (b) WC-inclusions 100–200 μm ϕ. (c) Fe-inclusions 500–800 μm ϕ.

The following advantages raise additionally high resolution radiography:

— analysis of greater cross-sections (cm^2) and volumes (cm^3) with microscopic resolution (μm);
— high depth of focus. Any detail inside a component will be imaged with the same U_g value;
— the projection technique increases the contrast by reduced scattered radiation ($I_S/I_D \rightarrow 0$);
— the primary magnification allows use of high sensitive film/screen combinations with short exposure times, otherwise applied only in medical diagnostics.

Fig. 14 gives some examples of microfocus radiography with image intensifiers.

It has already been mentioned that this technique also detects structural variations as a consequence of local density/absorption changes. But because of the projection of a three-dimensional object onto a two-dimensional screen, defect orientations and microstructure variations in different directions can only be found by multi-angle through-radiation, e.g. moving the object before the filmless operating receiver. This finally leads to X-ray tomography (see below) but, nevertheless, the relatively simple conventional projection technique is the main element of a quality control concept for ceramic components.

5.1.2 X-ray computer tomography (CT)

X-ray CT has belonged to the state-of-the-art in medical diagnostics for more than a decade. Its introduction in materials testing is still in an early stage but worldwide there are strong activities to bring this forward as an appropriate NDT technique. Referring to the necessarily high resolution required in analysing ceramics, it is just at the beginning.[36] During the next few years a breakthrough in 'Micro-CT', i.e. X-ray CT with μm-resolution, has to be expected. Results showing more macroscopic details by X-ray CT with conventional (medical) equipment are given in Fig. 15.

5.2 Ultrasonics

5.2.1 Defect testing

Ultrasonic testing and radiography cannot be considered as alternative techniques. In general, they complement one another:

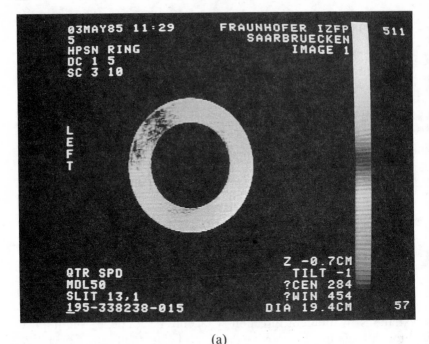

(a)

Fig. 15. X-ray computer tomography of defects in advanced ceramics.
(a) Inhomogeneities in a HPSN-ring. (b) Inclusion in a RBSN turbocharger
rotor; the rotor is embedded in Si powder to avoid artefacts. (c) Crack in the
axle of a SSiC-turbocharger rotor; again embedded in Si powder.

— a silicon inclusion in silicon nitride is hard to detect with X-rays
   because of a small difference in the absorption coefficient μ, while
   it comes out much clearer with ultrasound because of greater
   differences in sound impedance ρv;
— the defect orientation relative to the analysing beam is very
   important. Ultrasound reflects primarily defects perpendicular to
   the beam while X-rays detect them optimally if they are oriented
   parallel to the beam.
— Finally, the geometry of the components plays an important part,
   because ultrasound is primarily suited for more simple geometries
   while X-rays are optimal for complex shaped bodies.

This situation makes it necessary also to develop high resolution
ultrasonic techniques applying high frequency ultrasound.[37,38] The
goals are twofold: detection and interpretation of defects. While
detection can be achieved sufficiently obtaining defect signals with

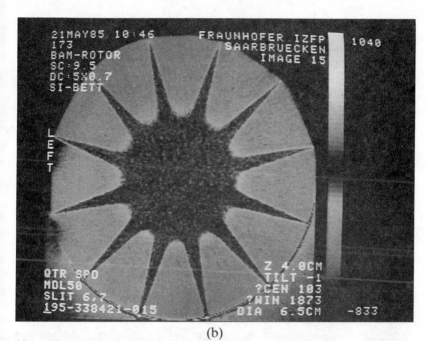

(b)

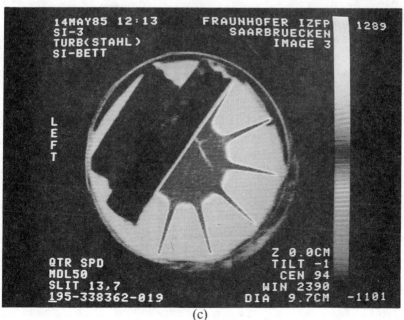

(c)

Fig. 15—*contd.*

amplitudes about twice the surrounding electrical or structural (scattering) noise, interpretation demands that these signals are received over a wide band of frequencies.

The interaction of a defect with an analysing ultrasonic wave is a function of the defect parameters (ρ, v, shape, size, orientation relative to the beam) and of the surrounding matrix parameters (ρ, v). This interaction can be calculated[39] and an experimental result can be compared to such calculations for defect identification.

According to Fig. 3 and Fig. 16 the interaction is typical for a given defect in the range $ka > 1$ ($k = 2\pi/\lambda, \lambda$ = wavelength, a = defect radius). Therefore, the so-called Rayleigh scattering ($ka < 1$) is good only for detection while the 'finger-print' of a defect starts above $ka = 1$. But this means that, disregarding a factor of π, the wavelength must be smaller than d (= $2a$) for identification purposes. Wavelengths of some $10\,\mu$m are obtained with frequencies of 100 MHz and more.

Two ways to realize this are possible:

— short pulses (pulse duration $\Delta \approx 1.5\lambda$) with a broadband spectrum[40]
— longer pulses ($\Delta \approx 5\lambda-20\lambda$) with a small frequency spectrum.[41]

Both methods have their advantages. At the moment, the second solution seems to be superior because of the higher dynamic gain reserve (amplitude of a backwall echo compared to the electrical noise) which is necessary to compensate for different types of losses (attenuation in the probe, loss in the coupling layer, attenuation in the ceramic, reflection losses at the defect, etc.).

Figure 17 sketches the instrument developed for this purpose[41] while different types of high-frequency ultrasonic probes are shown in Fig. 18. The system without the probes now has a dynamic range of 100 dB for the frequency range 50–200 MHz. An example for the frequency response of an unknown defect in hot-pressed silicon carbide is shown in Fig. 19. This finger-print correlates best with a spherical tungsten carbide-inclusion of 480 μm ϕ (X-ray image: 400 μm ϕ). The identification was correct (seeded tungsten carbide-particle in the object[41]). At the moment this technique is in a laboratory stage and has to be still further improved for application in practice.

5.2.2 Defect imaging

Acoustic microscopy resulting in two-dimensional images has already been described in Section 4.2. Applying frequencies no higher than 200 MHz especially in dense ceramics will make this method also

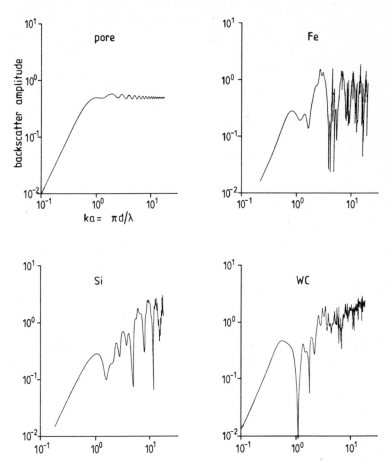

Fig. 16. Ultrasonic response of different types of spherical inclusions in HPSiC (calculated).

suitable for bulk material testing. At the moment, the manufacturers of acoustic microscopes with $f \approx 1$ GHz are redesigning their instruments for the lower frequencies. SLAM is able to directly image internal defects (30 MHz, 100 MHz) depending upon the attenuation and the thickness of the objects.

5.2.3 Ultrasonic stress measurement

The failure of ceramic parts is more and more discussed not with respect to defects but also regarding residual stresses. While X-rays are

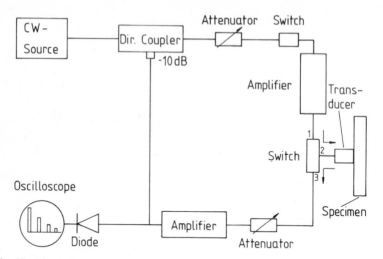

Fig. 17. Block diagram of electronic equipment assembled for high-frequency
ultrasonics.

able to detect and to quantify surface stresses, ultrasonic waves are able
to characterize surface stresses as well as volume stresses. While the
physical background was briefly indicated in Section 2.1, Fig. 20 gives
an example of the detection of residual stresses in a ceramic ring:
insonifying shear waves vertically with polarizations radially (rad) and

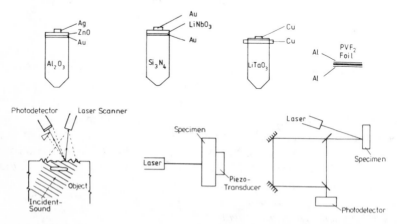

Fig. 18. Excitation and detection of high-frequency ultrasonic waves.

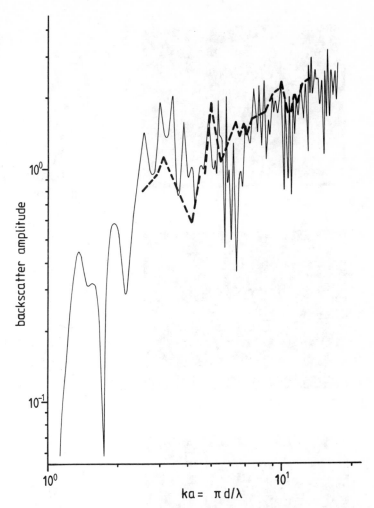

Fig. 19. Comparison of calculated (————) and measured (————) ultrasonic
    defect response, fitted to spherical WC inclusion of 400 μm diameter.

tangentially (tg) leads to relative velocity differences in the order of a few
tenths of a percent which are proportional to the stress difference
($\sigma_{rad} - \sigma_{tg}$). The quantitative figure of the stress difference cannot yet be
given because the third-order elastic constants of polycrystalline
zirconia are not known.

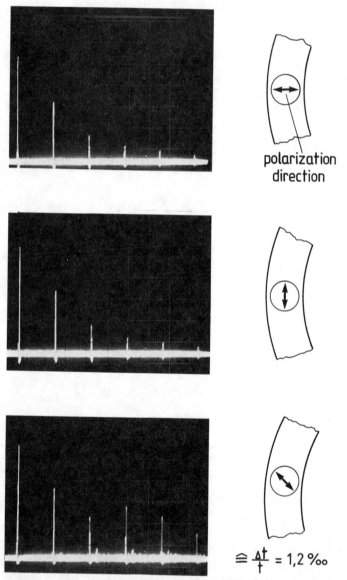

Fig. 20. Ultrasonic shear wave backwall echoes from a drawing tool (ZrO$_2$, thickness 64 mm, frequency 4·4 MHz) for different directions of polarization.

5.2.4 Guided ultrasonic waves

Guided ultrasonic waves (tube waves, plate waves, surface waves, etc.) are severely disturbed by defects as already mentioned for Rayleigh waves in Section 4.2. It depends therefore upon the component geometry if guided wave ultrasonic testing is appropriate or not. Up to now, one example of tube waves in silicon carbide heat exchanger tubes can be given: the excitation of the tube wave was obtained with an electromagnetic-acoustic transducer (EMAT) and the procedure was optimized by a 50 μm layer of copper along 10 cm at one end of the tube.[42] With a wavelength of 12 mm defects $\geqslant$3% of the wall thickness (5 mm) of the tube could be found, corresponding to a d/λ-value of $\approx$0·01.

6. INSPECTION OF BONDING LAYERS

Systems cannot be manufactured in one part. They generally have to be assembled by different components, very often made from different materials. The bonding of ceramics to metals or to other ceramics therefore is an important step during fabrication of systems and has to be accompanied by relevant quality control methods, i.e. NDT techniques. The special feature of bonding layers is the close contact between two surfaces which can be flat, curved or even extremely complicated like screw fittings. In any case, the bonding layer is very thin and the bonding can be realized with and without a bonding agent, e.g.

— shrinking of metals to ceramics,
— friction welding,
— diffusion bonding,
— cement- and slip-casting bond,
— brazing alloys.

The defects occurring are mainly debonded areas, pores, inclusions, cracks due to different heat expansion coefficients of the bonded parts. The discussion also has to include ceramic coatings on metal surfaces and metals sputtered/evaporated, etc., on to ceramic substrates. Here, besides the above mentioned types of defect, the measurement of coating thickness is an important NDT-goal.

6.1 Ultrasonic Testing

The different situations of ultrasonic testing can be discussed with reference to Fig. 21. The case of guided waves in coatings and bonding

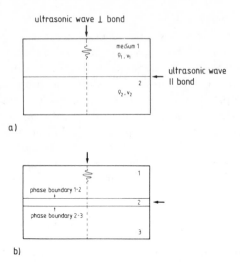

Fig. 21. Geometry of relevant bond types: (a) without bonding layer; (b) with bonding layer.[2]

layers is now physically well understood[43] but not yet experimentally realized for ceramic components. The usual testing arrangement is vertical incidence of pulses and echo amplitude evaluation.

Firstly regarding coatings and bonds without an intermediate adhesive layer, the signal amplitude from vertically incident waves will be proportional to the reflection coefficient R:

$$R = \Delta Z / \sum Z$$
$$\Delta Z = \rho_2 v_2 - \rho_1 v_1, \sum Z = \rho_2 v_2 + \rho_1 v_1, Z = \rho v = \text{sound impedance}$$

when the indices 1 and 2 indicate (refer to Fig. 21) the first and second medium.

The R values of some interesting ceramic–metal and ceramic–ceramic bonds can be calculated using the data in Table 3. For materials bonded to each other without a big ΔZ small amplitudes characterize good bonds while high amplitudes arise from badly or unbonded areas (total reflection, $R = -1$, in the worst case). An important fact is that R can take values $\lessgtr 0$ while $|R| \leqslant 1$, so the phase information (pulse sign) is a bond quality indicator too. Some examples for good and bad bonds of these types are given in Fig. 22. For thin layers between the bonded parts in most cases the axial resolution is not high enough to resolve

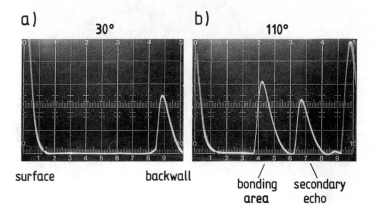

a) 30°   b) 110°

surface   backwall   bonding   secondary
                                        area   echo

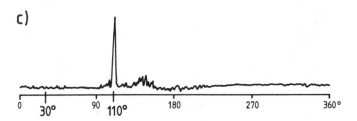

c)

0   30°   90 110°   180   270   360°

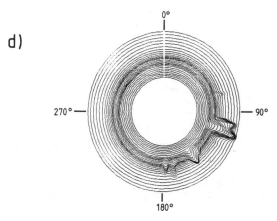

d)

0°

270°   90°

180°

Fig. 22. Ultrasonic testing of Al_2O_3-bond with metallic layer: (a, b) A-scans at
two different positions; (c) amplitude of bond-echo from 0° to 360°; (d) C-scan
of the bond layer.[18]

K. Goebbels

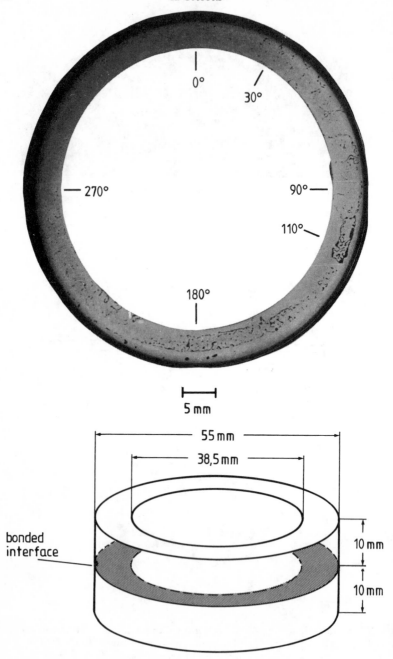

Fig. 23. Microradiography from Al$_2$O$_3$ bond (see Fig. 22).

echoes from the phase boundaries 1-2 and 2-3 (see Fig. 21). Then it is difficult to judge signal amplitudes because of the interference effects for the signals from the two phase boundaries.[44]

At the moment, ultrasonic testing of bonds is still under development and a real breakthrough cannot be foreseen immediately. Only for completeness it should be mentioned that acoustic microscopy is also a sensitive tool for quality control of coatings and bondings, if access is given, i.e. attenuation not too high, to the interface between the bonded materials.

6.2 Photothermal Microscopy

Photothermal microscopy has already shown its potential for bond quality assurance in the case of coatings:[31] plasma sprayed tungsten carbide layers and chromium oxide layers on steel were tested with a low frequency chopped laser beam (= 20 Hz) and a focussed infrared receiver. Corresponding to the ultrasonic case the reflection coefficient from the interface — if the penetration depth is sufficiently high — is given by:[31]

$$R' = \Delta Z'/\sum Z'$$
$$Z' = \sqrt{\rho c k}$$

with ρ = density, c = specific heat, k = thermal conductivity.

Here too, amplitude and phase of the signals contain the relevant information. Some examples for Z' and R' can be calculated using the data given in Table 5.

6.3 Radiography

Bond testing with radiography is problematic, because the direction of radiation should be in the plane of bonding. Otherwise only volumetric defects in the interface region can be detected.[44] Consideration of bonding along curved surfaces then leads directly to multiangle radiography and tomography. Figure 23 shows some examples from brazed joints (X-ray beam perpendicular to the bonding layer), where the correlation with the ultrasonic results (Fig. 22) is quite evident. In Fig. 24 the radiation direction was parallel to the bonding area of a HPSN rotor hub to the RBSN turbine blades. The inhomogeneous bond is seen from the variation in light intensity over the image.

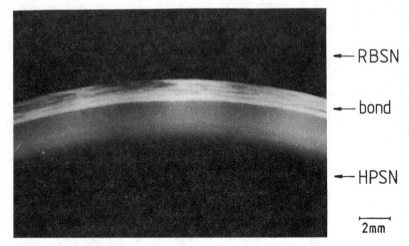

← RBSN

← bond

← HPSN

├─────┤
2mm

Fig. 24. Microfocus X-ray image of the bonding layer between HPSN hub and
RBSN blades of a ceramic gas turbine rotor.

7. ADDITIONAL NDE METHODS

7.1 Optical–Holographical Interferometry

With optical–holographical interferometry it is possible to image
defects in specimens under load (mechanical, thermal load). The ability
was shown by several experiments.[17, 34] But compared to UT and X-rays
the lack of resolution, especially for internal defects, does not
recommend this technique up to now.

From another point of view the scattering of focussed laser light at a
reflecting surface can be evaluated for surface characterization.
Especially for polished surfaces, e.g. ball bearings or bioceramic parts,
further analysis of this technique (small and wide angle scattering) can
be recommended. Experimental results are published for surface
roughness characterization of silicon nitride ball bearings.[45]

7.2 Thermography

The detection of defects by PTM has already been mentioned. With
much lower resolution conventional thermography can also be used. It
shows especially the integral quality of series of products, e.g. heat
exchanger tubes made from siliconized silicon carbide referring to their
heat conducting properties.[34] Thermography will play an increasingly
important role in NDT of integrated circuits and other electronic parts
and systems whereof ceramic substrates are relevant components.

7.3 Bubble Testing

Bubble testing is an adaptation of methods used for leak testing and has already found its application in the semiconductor industry to control seals of electronic packages. The method can be used for detection and characterization of surface connecting defects.

The advantage over conventional liquid penetrants is the possibility of quantitatively evaluating the number of emitted bubbles.

Parts to be tested are first suspended in an atmosphere of 15 kPa above a water bath for about 10 min to evacuate the defects. Then they have to be lowered into the water under slight hydrostatic pressure for another 10 min. Finally immersing them in hot oil at 125°C now emits the bubbles which can be analysed by measuring their diameter, counting their number and correlating this information with the volume where they come from. The first promising results were obtained for silicon carbide test bars.[46] But work to better judge the possibilities and the limits of bubble testing has still to be continued.

7.4 Magnetic Resonance Imaging

It was for green ceramics that nuclear magnetic resonance imaging (NMR) was used in the first experiments:[47] the idea that hydrogen-rich plastic binder material will fill green ceramic pores and that NMR then will image these volumes was not as successful as expected. But the detection of water in surface connecting defects in silicon carbide showed encouraging results.

The NMR technique, already in use in medical diagnostics (NMR tomography), still has to be further analysed with respect to their potential in materials NDE.

7.5 Acoustic Emission

Generally, the potential of acoustic emission analysis is very high for brittle materials. The elastic energy stored during a loading process (thermal, mechanical load) cannot be released by plastic deformation but will be used almost totally for the generation of cracks. Especially for oxide ceramics with grain sizes >10 μm this was proved in the past.[48] In agreement with the known literature, experiments have shown that for RBSN specimens with and without Knoop indentations, no correlations could be found with the strength or the damage.[17] The reason for this is assumed to be the fine structure with grain sizes <1 μm: too many 'bridges' are breaking individually instead of storing the elastic energy until the level of an 'audible' noise is reached. A complete judgement needs more detailed analysis of acoustic emission experiments.[49, 50] Promising results were reported on HPSN double torsion tests.[51, 52] The acoustic emission there could positively be related to the room temperature crack growing process.

7.6 Electrical and Electromagnetical Methods

Silicon carbide as well as siliconized silicon carbide or carbonized silicon carbide show limited electrical conductivity. Conductivity measurements[53] and eddy current testing with appropriate frequencies[42] make it possible to characterize the material and to detect defects. But development activities in these directions are very few.

By analogy to the eddy current testing of 'metallic' materials, microwave testing can be seen for materials without electrical conductivity. Especially in silicon nitride the semiconducting silicon as one of the most important types of defects can easily be detected. Detailed analysis has shown the microwaves' capability to detect other defects also like tungsten carbide, iron, boron nitride, graphite or pores.[54] The necessary frequency range of more than 100 GHz at the moment unfortunately includes disturbing signals from the component geometry (curvature, edges, varying thickness, etc.) which cannot be suppressed. Up to now, it seems not to be possible to develop this technique for practical application without extensive work.

7.7 Other Methods

As described for NMR and bubble testing, only some isolated results are known up to now for additional NDT methods. So krypton gas impregnation was reported[46] as a technique for filling surface connecting defects with [85]Kr gas and to expose a photographic emulsion to the emitted beta-radiation.

Another example was given for the analysis of microvoids (diameter 20 Å!) with small angle neutron scattering.[55] Here, for example, the thermal fatigue process of hot-pressed silicon nitride could be characterized nondestructively. In this context the method has to be considered as a very sensitive tool for basic materials characterization and for comparison with other NDT techniques.

High-temperature materials degradation was also studied with internal friction measurements.[56] The results are reported as coming from structural transformations especially in the (glassy) layers between the grains of sintered silicon nitride.

8. REMAINING NDE PROBLEMS

While Sections 3–7 describe developments in NDT regarding the state-of-the-art as well as problems still to be solved, this section now deals with questions already set but not yet treated with the necessary engagement.

8.1 'Green' Ceramics Testing

In the past most attention was given to ceramic materials in their final structural state. Only recently[47, 53] were some remarks made on the NDE of 'green' ceramics, i.e.

— differences in X-ray detectability for the same defect (boron carbide in silicon nitride) occurring in 'green' and in sintered ceramics due to the absorption coefficients in the different states;[53]
— elastic anisotropy of the 'green' state due to the fabrication process. This could be detected by polarized ultrasonic shear waves. The problem is the coupling;[47]
— high attenuation of ultrasound in green ceramics, e.g. 1–2 dB/mm at 2·25 MHz; coupling obtained by shear wave couplants or by direct contact between probe and specimen.[47]

From the economic point of view it is important to apply NDT methods to 'green' ceramics too, because for defective parts including inhomogeneous material distributions as well as single defects time consuming and expensive work can be avoided. The problems are, for example, that some of the 'green' ceramics are highly attenuating to ultrasound while others are very sensitive towards handling. In these cases too X-ray radiography seems to be the best choice.

8.2 Inspection of Complete Systems

After fabrication and eventually bonding of ceramic parts and components, they have to be integrated in full systems. For cold, and especially for hot, ceramic applications quality control problems arise here due to, for example, geometrical dimensions, eccentricity and imbalance. Up to now, no work has been done in this direction. First of all, vibration analysis of complete systems seems to be an appropriate way for NDE, including acoustic emission testing (under well defined loads), impact noise and modal analysis.

8.3 Service Induced Defects
The types of defects discussed so far in the preceding sections included only flaws, anisotropy and inhomogeneity due to manufacturing processes. Service induced defects, i.e. the occurrence of cracks in heavily mechanically[57] and thermally[58] loaded areas, fatigue and corrosion cracks, oxidation and other types of material degradation (creep) at high temperature are well known as failure causing effects. But their occurrence in real parts under real conditions as well as their nondestructive detection and evaluation are important problems which have to be solved in the future.

8.4 NDE-Based Fracture Mechanics
Accept/reject criteria for ceramic materials and components are based on stress analysis and fracture mechanics as shown in Fig. 1. But such analysis has to be made on the basis of NDE results, i.e. detection, location and interpretation (sizing, characterizing) of defects and heterogeneities. Up to now, the cooperation between NDT and fracture mechanics has not yet reached the intensive interaction of other areas. But referring to the small safety margin (see Section 1.2) between detection level and failure this is one of the most important questions to be worked on in the future if advanced ceramics are to find their place in the engineering of better and new components and systems.[59, 60] Here the cooperation between engineering and NDT has to be improved to combine:

— design for inspection, and
— inspection for quality assurance.

9. CONCLUSIONS

Advanced ceramics have several advantages which make them important materials for structural applications: low density, high strength and hardness, low heat expansion and good heat conductivity combined with high thermal stability and oxidation resistivity. But there is one disadvantage in that even at high temperatures advanced ceramics are brittle, which means that stress peaks at defects cannot be released by plastic flow and ductility. From the stress analysis and fracture mechanical point of view for design loads of some 100 MPa, defects with linear dimensions of 10–100 μm — at the surface, still smaller ones — will cause failure. This makes a nondestructive quality control absolutely necessary. Because conventional NDE methods are able to detect and to characterize defects with dimensions only about two orders of magnitude higher during the last decade special research and development efforts have been made to develop NDE techniques with high sensitivity and high resolution.

9.1 NDE Problems and NDE Methods

The NDE problems are, as in any other application, defects, structure and stress.

Surface defects are more critical than internal defects and cracklike defects are much more severe flaws than volumetric ones like inclusions. Therefore, NDE methods not only have to detect and to locate the defects, they also have to characterize them referring to their kind, size and orientation.

Microstructural features which have to be characterized by NDE are inhomogeneity and anisotropy. They will cause position-dependent and direction-dependent material properties.

Residual stresses finally can reduce the materials' ability to withstand the design loads, and their quantitative determination therefore means also a very important demand for NDE.

Special problems referring to defects, structure and stress are set by ceramic–ceramic and ceramic–metal bonds as well as by testing procedures for complete systems.

Following the priorities given above, NDE methods first of all have to cover the surface inspection problem. Visual inspection including the microscope as well as liquid (fluorescent) penetrants can be the starting point for qualitative NDT while dye-enhanced radiography and UT with surface waves are able to quantify the defects found. In the future acoustic microscopy as well as photoacoustic and photothermal microscopy will cover the surface inspection problem with the highest sensitivity and resolution.

Bulk material characterization can easily be achieved by high resolution radiography including multiangle testing and tomography. But ultrasonic testing with frequencies around 100 MHz will assist this type of inspection for types of defects and orientations of defects which are less appropriate for X-ray detection and interpretation.

Finally, from the mass production point of view vibration analysis in the form of simple resonance frequency testing as well as 'full' spectrum modal analysis will be a very attractive and economic quality control method before the more sophisticated NDE techniques given above are to be applied.

While the methods described so far are relevant for defect and for structure evaluation, stresses can be measured — residual stresses as well as load stresses — at the surface by X-rays as well as ultrasound and in the occurrence of internal bulk stresses by ultrasonic methods.

9.2 NDE Equipment

The full variety of NDT techniques described in the preceding sections has demonstrated their ability for quality control of advanced ceramics mostly in laboratory applications and in limited areas of applications. Only for high resolution X-ray testing in projection techniques does the state-of-the-art of the equipment allow a direct application in practice. But here the film is the up-to-now accepted receiver medium. The filmless technique using X-ray sensitive cameras, image intensifiers and other types of solid state detectors still needs further optimization.

For ultrasonic testing as well as for photoacoustics and vibration analysis the appropriate equipment is still under development. But the realization of apparatus and sensors applicable in practice under economic considerations can be foreseen for the next few years.

ACKNOWLEDGEMENTS

The research and development work described herein as far as it refers to IzfP was performed mainly with financial support of the German Ministry for Research and Technology (BMFT) in the scope of programmes for the development of ceramic gas turbines, while additional activities were financed by the Fraunhofer-Gesellschaft (FhG), by the European Community as well as by projects for different manufacturers of ceramic materials and components. Their support is gratefully acknowledged.

REFERENCES

1. Evans, A. G., Non-destructive failure prediction in ceramics. In: *Progress in nitrogen ceramics,* Riley, F. L. (ed.), Boston, Martinus Nijhoff, 1983, pp. 595–625.
2. Rosenfelder, O. and Reiter, H., *Fortschrittsberichte der DKG,* **1** (1) (1985) 103–12.

3. Marshall, D. B., Surface damage in ceramics. In: *Progress in nitrogen ceramics,* Riley, F. L. (ed.), Boston, Martinus Nijhoff, 1983, pp. 635–56.
4. Goebbels, K., Hirsekorn, S. and Willems, H., *1984 IEEE Ultras. Symp. Proc.,* pp. 841–6.
5. Schneider, E., Goebbels, K., Hübschen, G. and Salzburger, H. J., *1981 IEEE Ultras. Symp. Proc.,* pp. 956–9.
6. Goebbels, K. and Hirsekorn, S., *NDT International,* **17** (1984) 337–41.
7. Truell, R., Elbaum, C. and Chick, B. B., *Ultrasonic methods in solid state physics,* New York, Academic Press, 1969.
8. Goebbels, K., Structure analysis by scattered ultrasonic radiation. In: *Research techniques in NDT, Vol. IV,* Sharpe, R. S. (ed.), London, Academic Press, 1980, pp. 87–157.
9. Papadakis, E. P., *J. Acoust. Soc. Amer.,* **42** (1967) 1045–51.
10. Schmitz, V., Barbian, O. A., Gebhardt, W. and Salzburger, H. J., *Materialprüf.,* **27** (1985) 49–56.
11. Höller, P. In: *Proceedings International Symposium New Methods in NDT,* Berlin, DGZfP, 1979, pp. 13–20.
12. Sharpe, R. S and Parish, R. W., Engineering applications of microfocal radiography. In: *Microfocal radiography,* Ely, R. V. (ed.), London, Academic Press, 1980, pp. 43–81.
13. Reimers, P. and Goebbels, J., *Mater. Eval.,* **41** (1983) 732–7.
14. Berger, H. and Kupperman, D., *Mater. Eval.,* **43** (1985) 201–5.
15. Goebbels, K., Reiter, H., Hirsekorn, S. and Arnold, W., *Sci. Ceram.,* **12** (1983) 483–94.
16. Arons, R. M. and Kupperman, D. S., *Mater. Eval.,* **40** (1982) 1076–8.
17. Goebbels, K. and Reiter, H., ZfP-Verfahren für Komponenten der Keramik-Gasturbine. In: *Keramische Komponenten für Fahrzeuggasturbinen II,* Bunk, W. and Böhmer, M. (eds), Berlin, Springer, 1981, pp. 163–93.
18. Lindner, H. A., Thoma, H. J. and Müller, H. Research Reports 01ZC1833, Cremer Forschungsinstitut, Rödental, 1983 and 1984.
19. Xavier, C. and Hübner, H. W., *Sci. Ceram.,* **11** (1981) 495–502.
20. Weinstein, D. and Shaw, H. J., PVF_2-transducers for NDE of ceramics and brittle materials. Stanford University, Report AFOSR-TR 81-0881, Oct. 1981.
21. Tien, J. J. W., Khuri-Yakub, B. T., Kino, G. S., Marshall, D. B. and Evans, A. G., *J. NDE,* **2** (1981) 219–29.
22. Fahr, A., Johar, S. and Murthy, M. K., *Review of progress in quantitative NDE, Vol. 3A,* Thompson, D. O. and Chimenti, D. E. (eds), New York, Plenum Press, 1984, pp. 239–49.
23. Kupperman, D. S., Caines, M. J. and Winiecki, A., *Mater. Eval.,* **40** (1982) 774–82.
24. Kessler, L. W. and Yuhas, D. E., *Scann. El. Microscopy,* **1** (1978) 555–60.
25. Lemons, R. A. and Quate, C. F., Acoustic microscopy. In: *Physical acoustics, Vol. XIV,* Mason, W. P. and Thurston, R. N. (eds), New York, Academic Press, 1979, pp. 1–92.
26. Reiter, H. and Arnold, W., *Beitr. Elektronenmikroskop. Direktabb. Oberfl. (BEDO),* **17** (1984) 129–36.
27. Rosencwaig, A., Thermal-wave imaging and microscopy. In: *Scanned image*

microscopy, Ash, E. A. (ed.), London, Academic Press, 1980, pp. 291–317.

28. Ash, E. (ed.), *Scanned image microscopy*, London, Academic Press, 1980.

29. Goebbels, K. and Reiter, H., IzfP-Res. Report 810127, Saarbrücken, 1981.

30. Wong, Y. H., Scanning photo-acoustic microscopy. In: *Scanned image microscopy*, Ash, E. A. (ed.), London, Academic Press, 1980, pp. 247–71.

31. Jaarinen, J. and Luukkala, M., In: *Proc. 3rd European Conference on NDT*, Italian Society for NDT, Florence, Italy, 1984, pp. 128–138.

32. Khandelwal, P. K., Heitman, P. W., Silversmitz, A. J. and Wakefield, T. D., *Appl. Phys. Lett.*, **37** (1980) 779–81.

33. Kleer, G. Richter, H., Prümmer, R. and Pfeiffer-Vollmar, H.-W., In: *Keramische Komponenten für Fahrzeug-Gasturbinen III*, Bunk, W., Böhmer, M. and Kißler, H. (eds), Berlin, Springer, 1984, pp. 487–512.

34. Kupperman, D. S., Sather, A., Lapinski, N. P., Sciammarella, C. and Yuhas, D., In: *Review of progress in quantitative NDE 1978*, Thompson, D. O. (ed.), Air Force Materials Laboratory Report AFML-TR-78-205, January 1979, 214–27.

35. Knoll, G. F., *Radiation detection and measurement*, John Wiley, New York, 1979.

36. Kress, J. W. and Feldkamp, L. A., *X-ray tomography applied to NDE of ceramics*, ASME-Paper 83-GT-206, 1983.

37. Tittmann, B. R., Ahlberg, L., Evans, A. G., Elsley, R. K. and Khuri-Yakub, B. T., *1976 IEEE Ultras. Symp. Proc.*, pp. 653–8.

38. Kino, G., Khuri-Yakub, B. T., Murakami, Y. and Yu, K. H., *Proceedings Review of Progress in Quantitative NDE 1978*, Thompson, D. O. (ed.), AFML-TR-78-205, 1979, pp. 242–5.

39. Chou, C. H., Khuri-Yakub, B. T., Kino, G. S. and Evans, A. G., *J. NDE*, **1** (1980) 235–47.

40. Khuri-Yakub, B. T. and Kino, G. S., *1976 IEEE Ultras. Symp. Proc.*, pp. 564–6.

41. Goebbels, K., Reiter, H., Arnold, W. and Hirsekorn, S., In: *Keramische Komponenten für Fahrzeug-Gasturbinen III*, Bunk, W., Böhmer, M. and Kißler, H. (eds), Berlin, Springer, 1984, pp. 537–58.

42. Reiter, H., Becker, R. and Coen, G., IzfP-Res. Report 800121, 1980.

43. Rokhlin, S. I., Hefets, M. and Rosen, M., *J. Appl. Phys.*, **52** (1981) 2847–51.

44. Reiter, H., Arnold, W. and Goebbels, K., *Proc. 2nd Intern. Colloq. Joining of Ceramics, Glass and Metal*, Bad Nauheim, March 1985.

45. Roszhart, T. V., *Holographic characterization of ceramics*, TRW-Report AD 776454, Redondo Beach, CA, TRW, 1973.

46. Friedman, W. D., *Review of Progress in Quantitative NDE*, Plenum Press, New York, 1984.

47. Kupperman, D. S., Ellingson, W. A. and Berger, H., Proc. NSF/JSPS US/Japan Seminar on Structural Ceramics, 1984.

48. Graham, L. J. and Alers, G. A., Microstructural aspects of acoustic emission generation in ceramics. In: *Fracture mechanics of ceramics*, Vol. 1, Bradt, R. C., Hasselman, D. P. H. and Lange, F. F. (eds), New York, Plenum Press, 1984, 175–88.

49. Schuldies, J. J. *Mater. Eval.*, **30** (1973) 209–13.

50. Gogotsi, G. A., Kuzmenko, V. A., Grishakov, S. V., Negovski, A. N. and Drozdov, A. V., Proc. 10th WCNDT, USSR Academy of Sciences, Moscow 1982, Paper 5-13.
51. Iwasaki, H. and Izumi, M., *Jap. Soc. Mater. Sci.*, Japan (1981) 1044-50.
52. Moeller, H. H., Powers, T., Petrak, D. R. and Coulter, J. E., *Proceedings Failure Mechanisms in High Performance Materials*, National Bureau of Standards, Washington, May 1984.
53. Grellner, W., Henze, P. Schwetz, K. A. and Lipp, A., *Proceedings 3rd European Conference on NDT*, Italian Society for NDT, Florence, Italy, October 1983, pp. 78-87.
54. Bahr, A. J., Microwave techniques for NDE of ceramics. In: *Review of progress in quantitative NDE 1977*, Thompson, D. O. (ed.), AFML-TR-78-55, May 1978, pp. 245-50.
55. Pizzi, P., Analysis of microvoids in Si_3N_4 ceramics by small angle neutron scattering. *Proc. Penn. State Conf.*, Pennsylvania State University, 1978, pp. 85-98.
56. Bonetti, E., Evangelista, E. and Gondi, P., In: *Internal friction and ultrasonic attenuation in solids*, Smith, C. C. (ed.), Oxford, Pergamon Press, 1980, pp. 401-5.
57. Khuri-Yakub, B. T., Shui, Y., Kino, G. S., Marshall, D. B. and Evans, A. G. *Review of Progress in Quantitative NDE, Vol. 3A*, Thompson, D. O. and Chimenti, D. E. (eds), New York, Plenum Press, 1984, pp. 229-37.
58. Ouanezar, N., Rouby, D., Fleischmann, P. and Fantozzi, G., *Sci. Ceram.*, **12** (1984) 563-8.
59. Richardson, J. M. and Evans, A. G., *J. NDE,* **1** (1980) 37-52.
60. *Proceedings Review of Progress in Quantitative NDE 1979*, Thompson, D. O. and Thompson, R. B., AFWAL-TR-80-4078, July 1980, Session XIV: Failure Modes, Defect Characterization and Accept/Reject Criteria.

7

Non-Oxide Technical Ceramics

Elektroschmelzwerk Kempten GmbH, Kempten, FRG

ABSTRACT

An overview is presented of the present status of the advanced non-oxide technical ceramics, namely borides, carbides and nitrides, including an assessment of their present effectiveness, reliability, and future potential. Factors influencing cost and commercial competitiveness and which lead to future research and development requirements, are mentioned. Many examples of application are discussed, with emphasis on non-engine parts.

1. INTRODUCTION

Ceramic materials based on borides, carbides and nitrides are generally grouped under the heading of non-oxide special ceramics. Of the inorganic non-metallic materials, this group stands out against others, since its constituent compounds do not exist as naturally occurring raw materials which can be processed directly. In contrast, they have to be synthesized with the aid of thermochemical processes, using metal oxides, boron, carbon and nitrogen as starting materials. Although the basic chemical compounds have been known for many years, their development into technically useful engineering materials is the result of intensive research and development activities, predominantly during the last two decades.

The non-oxide special ceramics have properties which are of great advantage for applications in many technical areas. Typical examples

151

of their properties are high hardness, high strength at low and high temperature, wear resistance, corrosion resistance, and thermal and electrical conductance, or resistivity.[1,2] In order to exploit the special properties of these materials in a real component, a substantial problem has commonly to be overcome, that of machining and finishing the component to close tolerances and high quality surface finish.

As shown in Fig. 1, borides, carbides and nitrides belong to the hardest available materials. Machining can only be carried out by diamond grinding techniques.[3] Additionally, the low fracture toughness of these materials (typically 3–7 MPa m$^{1/2}$, against the hard metals 25 MPa m$^{1/2}$ for comparison) causes further difficulties. They all show brittle fracture behaviour.

Looking at the properties and applications of the non-oxide special ceramics, the necessary machining of these materials should not be forgotten. Possible technical advances are often closely connected to the question of whether or not the necessary machining can be mastered in a commercially competitive way.*

The following examples of application concentrate on existing industrial solutions; future applications and development trends are mentioned.

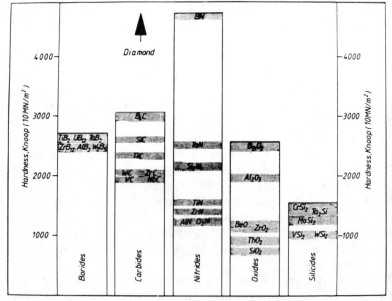

Fig. 1. Knoop-100 hardness of various hard materials.

2. BORIDES

The market for pure boride ceramics is comparatively small. Nevertheless they have extensive potential applications for the future.[4] The diborides TiB_2 and ZrB_2, and the hexaborides EuB_6 and LaB_6, are of special importance. Diborides show an excellent chemical stability, high melting point (T_m TiB_2 $\approx$ 3200 °C) and metallic electrical conductance.

The stability of TiB_2 in contact with cryolite, and aluminium at 1000° C suggests this material for the construction of cathodes in the aluminium smelting process. By replacing conventional graphite cathodes by TiB_2, the anode–cathode interpolar distance in a smelting cell can be reduced, since the TiB_2 surface is wetted by liquid Al. As a consequence ohmic losses in the bath can be reduced, giving energy savings of up to 25%. Sintered diboride crucibles can be used to handle liquid metals on a laboratory scale, e.g. Mg, Al, Cu, Ag, Au, Zn, Cd, Ge, Pb, Bi and Cr. The exceptionally high Young's modulus of TiB_2 (550 GPa) is useful for the production of armour plate, especially for protection against high speed penetrating projectiles.

Hexaborides show a low work function for electrons, 2·68 eV in the case of LaB_6. It is therefore a useful material for electron emission cathodes, able to achieve high current densities at comparatively low temperatures. Recent research results indicate that sintered LaB_6 is a promising construction material for the hydrogen-fluorine chemical laser.[5]

All the metal borides can be used as neutron absorbers, since the ^{10}B isotope (19·9 at.% of natural boron) has a high capture cross-section for neutrons. If the metal of the metal boride is also a neutron capturing element (e.g. the lanthanides), then the neutron absorbing properties of the boride compound become even better. EuB_6 for example is being tested in neutron flux control rods of fast flux breeder reactors.[6] Figure 2 shows sintered EuB_6 pellets for control rods. The neutron absorbing ^{10}B isotope can be enriched to a level of more than 90 at.% of the total boron, so increasing the efficiency of the absorbing material.

3. CARBIDES

Technically the most important ceramic carbide materials are silicon carbide (SiC) and boron carbide (B_4C). The first SiC ceramic on the market was silicon-infiltrated, or reaction sintered, SiC. This material is

Fig. 2. EuB$_6$ pellets, neutron absorber for fast neutron flux control rods.

a composite of SiC (85–90 wt %) and metallic Si (10–15 wt %). It is only since the late 1970s that pure sintered SiC has become commercially available. The SiC materials exhibit excellent corrosion resistance, high hardness, high thermal conductivity and thermal shock resistance.[7] The first major application of SiC was in mechanical seal rings, Fig. 3. The excellent wear resistance and tribological performance[8] of SiC guaranteed a longer life for the seal, thus reducing maintenance and production costs in the chemical industry, where pumps equipped with such seals are widely applied.[9]

During the last few years there has been a strong tendency in the chemical industry to use the pure sintered SiC, since Si-infiltrated SiC is chemically attacked and corrosively destroyed by some acids and in alkaline solutions. Pure sintered SiC in contrast shows a universal stability to all acids, alkaline solutions, and organic solvents, and simultaneously exhibits excellent abrasion resistance. These facts are the background for the outstanding success of media lubricated sliding bearings in hermetically sealed pumps, Fig. 4.

Fig. 3. Pure sintered SiC mechanical seal rings.

Only by using pure sintered SiC as the bearing material, could magnetically driven pumps be universally applied. Even with corrosive, and hard particle containing media, the life performance of pure sintered SiC sliding bearings is outstanding. Other pump components, exploiting the wear resistance of SiC, are rotating shaft sleeves, or precision spheres for dosing valves, Fig. 5. Nozzles for spraying systems take advantage of the corrosive and abrasive wear resistance of pure

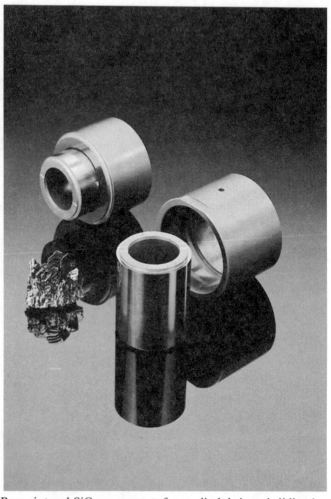

Fig. 4. Pure sintered SiC components for media lubricated sliding bearings.

sintered SiC; they are used for example, in desulphurization plants in coal or oil fired power stations.

Another application, demonstrating the corrosion resistance and widespread application potential of pure sintered SiC, is found in the form of stirrers, which are used to homogenize Al/Si-melts in the production of semiconductor grade polycrystalline Si.

Besides their favourable mechanical and corrosive wear properties,

Fig. 5. Pure sintered SiC spheres.

SiC materials show an excellent potential for high temperature applications. Applications range from burners to heat exchangers (Fig. 6), to the huge field of automotive applications.[10] Publications in this field are widely known, the 'ceramic motor' is a common head-line of our time.

If, and to what extent, commercial success can be achieved in this field, is a question which will be answered by the research and

Fig. 6. Recuperative burner, containing SiC and Si_3N_4 components.

development work of the next few years. A component showing the advanced state of development is shown in Fig. 7. Turbocharger rotors of this type have been successfully tested under service conditions.

Boron carbide is best recognized for its hardness and abrasion resistance.[4] After diamond and cubic boron nitride, B_4C is the third hardest of the technically useful materials. Sand blasting nozzles made of dense sintered B_4C are extremely wear resistant (Fig. 8). Under highly abrasive conditions B₄C outperforms other hard materials. If hard alumina grits are used as blasting media for example, then the life-time of a B_4C nozzle is about 100 times better than a hard metal WC nozzle under the same blasting conditions (Fig. 9). The extraordinary hardness of B_4C can be exploited also in the form of dressing sticks to contour grinding wheels; this means that sintered B_4C can be used to machine grinding wheels (it cannot be used for diamond wheels, however, which

Fig. 7. Pure sintered SiC turbocharger rotors.

Fig. 8. Sintered boron carbide sand blasting nozzles.

are used to machine B_4C). Wear resistant B_4C mortars or grinding media are used in trace chemical analysis, when pick-up of impurities has to be avoided. This is especially important if the material to be analysed is of an abrasive nature.

The low density of B_4C ($2 \cdot 51$ g/cm^3) and its high Young's modulus (440 GPa) recommend this material for the construction of light-weight armour, as is needed in the military helicopter and similar aero applications.

B_4C can also be used advantageously to control the neutron flux of nuclear reactors. By enriching the ^{10}B isotope from 19·9 at.% of the natural value to concentrations of above 90 at.%, neutron absorbers of different efficiencies can be produced. B_4C pellets, similar to those

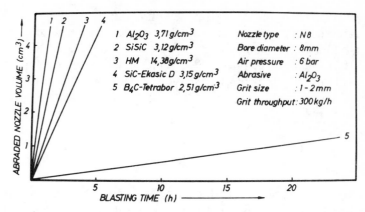

Fig. 9. Comparison of different hard materials in a wear test as sand blasting nozzle.

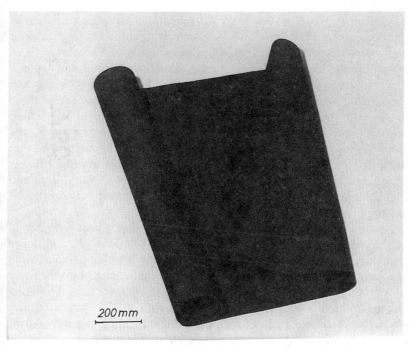

Fig. 10. Flexible foil, containing B₄C for neutron absorption.

shown in Fig. 2, are normally stacked into neutron control rods. B_4C materials are used to control neutron fluxes in boiling water, pressurized water, high temperature, and fast flux breeder, reactors. In addition to applications in the reactor core, B_4C based materials (plates, foils, etc.) are used to absorb neutron radiation in, for example, laboratory experiments, and during transport and storage of neutron radiating materials (Fig. 10).

4. NITRIDES

Among the ceramic nitride materials, those based on aluminium nitride (AlN), boron nitride (BN), and silicon nitride (Si_3N_4) have gained technical importance. While borides exhibit metallic character, with regard to electrical properties, the carbides show semiconductor properties; the nitrides listed above are electrical insulators.

Pure hexagonal BN is a soft material (like graphite) and retains its excellent electrical insulating properties up to temperatures above 2000° C. It is therefore used for thermocouple protection tubes and electronic insulating components. However, it can only be used as a high temperature electrically insulating material under inert atmospheres or vacuum conditions; it oxidizes in air or oxygen-containing atmospheres. A further conspicuous property of BN is that it is not wetted by molten glass, metals and slags; it shows excellent corrosion resistance in these media. The thermal shock resistance of BN is very good, too. These properties are exploited in the application of BN as break rings for the continuous horizontal casting of metals.[11] It should be mentioned that the commercial breakthrough of the horizontal casting of steel was only achieved by the use of BN rings. Figure 11 shows such rings, and Fig. 12 shows the schematic arrangement of a BN ring in the continuous casting technique.

Aluminium nitride (AlN) shows an exceptional combination of properties: high thermal conductivity (140–160 W m^{-1} K^{-1} at room temperature) and a high electrical resistivity (10^{12}–10^{14} ohm cm at room temperature), which decreases very slowly with temperature. These properties make AlN a favourable substrate material for high power electronic modules. In these applications AlN will replace beryllium oxide (BeO) substrates. BeO is an excellent substrate material in terms of physical properties. However, its extreme toxic effects are a great handicap in production and application.

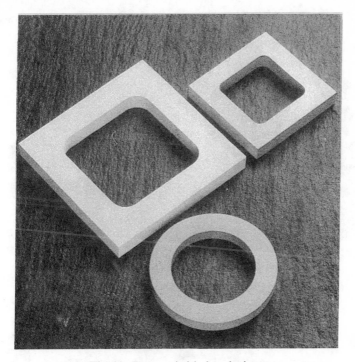

Fig. 11. Boron nitride break rings.

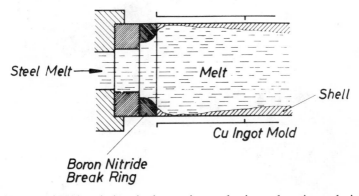

Fig. 12. Use of BN break rings in the continuous horizontal casting technique.

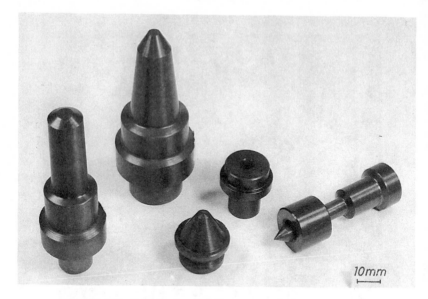

Fig. 13. Dense sintered Si_3N_4 valve components.

Fig. 14. Dense sintered Si_3N_4 ball-bearing components.

The corrosion resistance of AlN in liquid aluminium is excellent. Therefore it is used for crucibles when melts of high purity have to be handled, for example Al/Si melts in the semiconductor industry.

Silicon nitride materials immediately come to mind, in the context of the ceramic motor and the ceramic high temperature gas turbine. As already noted with SiC, this technical field defines a future market. For the engineering components of today, wear resistance is the most important property. Among the non-oxide ceramic materials the dense Si_3N_4 materials have highest strength and the highest fracture toughness. With regard to cavitation resistance all other materials are outperformed. In this field Si_3N_4 materials are the only realistic solution when high pressure relief valves have to be used at temperatures of some hundred degrees centigrade, as for example in coal liquidification and gasification. Production of these components demands excellent machining skills. Figure 13 shows some valve components.

The high strength and fracture toughness, accompanied by good corrosion resistance, is being exploited in the construction of Si_3N_4 ball-bearings, Fig. 14, which can be used in corrosive environments and at raised temperatures, where lubrification becomes a problem. Today, Si_3N_4 ball-bearings demonstrate their problem-solving potential, especially in the chemical industry, in, for example, long-lasting reliable bearings for process stirrers.

For the cutting of metals, especially for cast iron, Si_3N_4 cutting tools will dominate the market[12] (Fig. 15). The cutting efficiency and extended

Fig. 15. Dense sintered Si_3N_4 cutting tools.

life of the cutting tool lead to savings in the production line, accompanied by a more secure level of production. The development of tailor-made Si_3N_4 materials is creating a steadily growing application base for cutting tools. Even the cutting of superalloys and highly alloyed steels will be handled by silicon nitride materials. For wire-drawing purposes, Si_3N_4 nozzles operate successfully.

Dense Si_3N_4 materials contain small amounts of a second phase, which is added as a sintering aid during production of the material. Reaction bonded Si_3N_4 (RBSN), in contrast, is of high purity, but contains about 15–20 vol.% of porosity. This material is advantageously used in applications where chemical stability is the crucial property. RBSN is stable in many non-ferrous metal melts, and is used for example as a crucible material in the production of semiconductor silicon. In the semiconductor industry many electrically insulating components are made of RBSN.

The application potential of Si_3N_4 materials is a large one. So far, however, the solutions of wear problems provide the dominant applications.

5. COMPOSITE MATERIALS

Combining the properties of the non-oxide special ceramics has often been tried, in order to achieve tailor-made properties for special applications. The best success has been achieved with the so-called evaporator materials, composites of TiB_2, BN and AlN. These materials are machined into 'boat' geometry and heated by direct electrical current in vacuum to temperatures of about 1450° C. Al wire fed continuously into the boat, Fig. 16, firstly melts and wets the boat material, and subsequently evaporates. Al coatings for capacitor foils, wrapping foils, mirrors, etc., are produced by this technique.

The special demands which are placed on an evaporator boat are exact electrical resistance, thermal shock resistance, and corrosion resistance against liquid Al at 1500° C.

6. CONCLUSION—FUTURE REQUIREMENTS

Non-oxide technical ceramics have proved their quality and reliability in industrial processes, when compared to other materials. They are recognized as typical problem-solving materials, i.e. they are used when

Fig. 16. Glowing evaporation boat with feed of Al wire.

other materials fail. In spite of this very positive situation — many industrial processes crucially depend on the performance of non-oxide ceramics — the worldwide production of non-oxide ceramics is still very small, when compared with oxide ceramics used for example for refractories, electronic substrates, magnetic and electrical applications. The situation will change dramatically, if the huge application potential for automotive uses can be exploited.

In order to be successful and to increase the market impact of non-oxide technical ceramics in the future, two important demands have to be met. The first is a scientific requirement; the materials have to be improved. Depending on the application, the demands on the properties are different. Strength and fracture toughness need to be increased. Ceramic/ceramic or ceramic/metal joints are still a problem, especially at high temperatures. The tribology of ceramic systems will become something more than an empirical science. In the future, materials science in this field will include research on multiphase ceramic systems. Reinforcement by second phases, such as fibres and whiskers, will be equally important as the research on ceramic alloys. Research on improved raw materials must continue.

The second demand lies in the field of process technology. In order to substitute an existing material by a better one, not only the properties but also the price, are of major importance. Competitive prices are the result of competitive fabrication technology. Further improvements in production technology, from the powder to final machining, are a steady requirement for producers. In this field not only the material scientist, but also chemists, physicists and process engineers, will have to master many tasks.

REFERENCES

1. Grellner, W., Schwetz, K. A. and Lipp, A. Properties and applications of special ceramics (German), *Radex-Rundschau,* **1/2** (1983) 146–51.
2. Knoch, H. *Carbide and boride ceramics — availability and commercialization*, Proceedings of the 1st European Symposium on Engineering Ceramics, London, 1985, pp. 77–91, Oyez Scientific and Technical Services Ltd.
3. Kessel, H. and Gugel, E. Parameters influencing the machining of hot-pressed Si_3N_4 components with diamond tools (German), *IDR,* **3** (1978) 180–5.
4. Schwetz, K. A., Reinmuth, K. and Lipp, A. Processing and industrial applications of refractory boron compounds (German), *Radex-Rundschau,* **3** (1981) 568–85.
5. Sheppard, L. M. Hot new applications for ceramics, *Advanced Materials and Processes,* **II** (1985) 39–43.
6. Reinmuth, K., Lipp, A., Knoch, H. and Schwetz, K. A. Boron containing neutron absorber materials (German), *Journal of Nuclear Materials,* **124** (1984) 175–84.
7. Hunold, K., Knoch, H. and Lipp, A. Processing and properties of silicon carbide components (German), *Sprechsaal,* **116** (1983) 158–62.
8. Knoch, H. and Kracker, J. Sintered silicon carbide, a material for corrosion and wear-resistant components in sliding applications, *cfi/Ber. DKG,* **64** (1987) 159–63.
9. Knoch, H., Kracker, J. and Schelken, A. Silicon carbide — material for erosion and corrosion resistant pump components (German), *Chemie, Anlagen, Verfahren,* **3** (1985) 101–4.
10. Bunk, W. and Böhmer, M. *Keramische Komponenten für Fahrzeug-Gasturbinen,* Berlin, Heidelberg, New York, Tokyo, Springer, 1984.

11. Hunold, K. and Sindlhauser, P. Properties and applications of boron nitride components (German), *Metall,* **10** (1985) 908–10.
12. Steinmann, D. Silicon nitride ceramics for cutting tools, today and tomorrow (German), *Sprechsaal — Fachberichte für Metallbearbeitung,* **62** (1985) 635–42.

8

Ceramic Matrix Composites

B. CALES

Céramiques Techniques Desmarquest, Trappes, France

ABSTRACT

Advanced ceramics such as alumina, zirconia, silicon nitride and silicon carbide are characterised by good resistance to wear, oxidation and corrosion, when compared with metals and thermoplastics. However, the use of monolithic ceramics is often limited by their low mechanical reliability. Ceramic matrix composites, with refractory particles or fibres dispersed as a second phase in a ceramic matrix, are characterised by a higher degree of mechanical reliability, and may be the subject of industrial development for specific applications. The aim of this contribution is to give a survey of ceramic matrix composites, summarising the state of the art in their manufacture, their physical properties, and their potential industrial development in the near future. Special attention will be given to whisker reinforced ceramics that appear to be a very promising class of composites because of their good mechanical properties and their simple manufacturing routes.

1. INTRODUCTION

Advanced ceramics such as alumina, zirconia, silicon nitride and silicon carbide are characterised by good resistance to wear, oxidation and corrosion, when compared with metals and thermoplastics. They have been developed for various applications in metallurgy, in heat engine systems, and the aerospace industries.

However, the use of ceramic materials is often limited by their brittleness, that is their tendency to fail catastrophically by the growth of

171

a single crack that originates from a small flaw. Monolithic ceramics are characterised by relatively low fracture toughness, and probabilistic fracture strengths determined by an inherent flaw population. In the last few years, many studies have been focused on the improvement of the mechanical reliability of monolithic ceramics, mainly consisting in reducing flaw populations, and optimising microstructures. For silicon nitride ceramics for instance, the refinement of grain size and shape to obtain fine grained microstructures with elongated fibrous-type grains has been carried out, and has led to a toughness increase by a factor of ~2.[1]

Recently, large toughness improvements have been obtained by designing new microstructures that consist of refractory particles or fibres dispersed as a second phase in a ceramic matrix. The performances of the early ceramic composites remained quite limited, compared to monolithic ceramics.[2] The origins of such limitations were essentially the large dimensions of the reinforcing phase, and its clustering due to processing difficulties.[2] It was shown later that significant toughness improvements require, firstly, that the diameter of the fibres or particles should be smaller than the typical flaw size in the ceramic matrix ($\sim$20–50 μm) and, secondly, that the second phase should be dense and uniform in order to reduce the inter-particle or inter-fibre distance to a dimension smaller than the typical flaw sizes of the matrix.[2]

The recent development of submicronic ceramic powders and very fine refractory fibres, related to improvements in ceramic manufacturing processes, explains the resurgence of interest in ceramic matrix composites. Thus several specific issues have been brought into focus on this topic.[3-5]

This paper will give a survey of ceramic matrix composites with respect to the state-of-the-art in their manufacture, their physical properties, and their potential industrial development in the near future. Special attention will be paid to whisker reinforced ceramics that appear to be a very promising class of composites because of their good mechanical properties and their simple manufacturing routes.

For reasons of clarity, the composites will be denoted as follows: Matrix/Dispersoid.

2. BACKGROUND

There is considerable interest in ceramic matrix composites because of their various potential applications. The physical properties of the ceramic matrix can be modified for the purpose of a specific application by adding an appropriate second phase, leading to an enhancement of either the electrical properties or the wear behaviour for instance. Nevertheless, the most common motivation for the development of ceramic matrix composites originates in the possibility of an increase in mechanical reliability resulting from an improvement of the mechanical characteristics of ceramic composites, especially fracture toughness. A number of mechanisms seems to contribute to ceramic composite toughening, such as: *load transfer, prestressing, microcracking, phase transformation, crack impediment* or *deflection, fibre pull-out, crack bridging.* The purpose of this paper is not to report detailed theories about these different ceramic toughening mechanisms that have been either analysed or modelled in several reviews.[2, 6, 7]

2.1 Load Transfer

Most of these mechanisms operate in the presence of stress fields along the matrix/dispersoid interfaces due to thermal expansion and/or elastic modulus mismatches. This is so in the case of toughening due to load transfer which arises in fibre-reinforced ceramics from an elastic modulus mismatch between the matrix and the fibres. Load transfer from the matrix to the fibres requires that the Young's modulus of the fibre (E_f) be greater than that of the matrix (E_m) and a ratio $(E_f)/(E_m) > 2$ would be more favourable.[8] This mechanism is more efficient in the case of highly organised long fibres rather than for randomly distributed short fibres or whiskers.[2] It is a primary factor for toughening in like/like composites, such as carbon/carbon or silicon carbide/silicon carbide composites, and for long fibres, that is carbon or silicon carbide reinforced glass and glass-ceramic composites.

2.2 Prestressing

Prestressing of the matrix or dispersoids arises from thermal expansion mismatches between the matrix and the dispersoids that lead, during cooling from the processing temperature, to either compressive or tensile stresses in the matrix. It is a very common mechanism, encountered in most ceramic composites provided that the nature of the dispersoid is different from that of the matrix. Like/like composites thus do not fall in this class.

2.3 Microcracking

Thermal expansion mismatches also can involve matrix microcracking that can contribute to toughening by means of a crack branching phenomenon. Detailed analysis of microcracking toughening has been reported by Rice[2] and by Coyle *et al.*[45] Strict requirements for microstructural dimension and uniformity are needed for effective microcracking.[9, 10]

2.4 Phase Transformation

Phase transformation may also be an important factor for microcracking. However, phase transformation toughening in practical systems mainly originates from the tetragonal to monoclinic stress-induced transformation of zirconia particles in a limited zone of stress concentration near the crack tip.[12] Modelling of zirconia-toughening has been developed by various authors.[10, 13, 14] The toughening arises from the volume expansion of the zirconia particles during the transformation from the tetragonal phase to the monoclinic that causes a relaxation of the matrix stresses near the crack tip. A size effect has been observed[14] so that constrained zirconia particles below a critical size retain the metastable tetragonal microstructure during cooling from processing temperatures, whereas particles larger than $\sim 1\,\mu$m transform spontaneously to the monoclinic phase despite the matrix constraint, and lead to the microcracking phenomena.[14] The transformation of near-surface zirconia particles during machining of zirconia-toughened ceramics[15] results in compressive stresses on the surface that also can lead to a strengthening of the material. Zirconia-toughened alumina is a current ceramic composite that relies on the transformation toughening mechanism and it has been the subject of numerous studies.[16, 17] Silicon nitride ceramics have also been reinforced by means of zirconia particles.[18]

2.5 Crack Impediment and Deflection

These mechanisms require a fracture-resistant second phase, such as particles or fibres, and are highly dependent on the dispersoid–matrix property mismatch.[7, 19] The initial crack can be deflected by sufficiently weakened dispersoid–matrix interfaces which are preferred crack paths. A typical microstructure of titanium carbide reinforced silicon nitride composite resulting from a crack deflection mechanism is shown in Fig. 1.

The crack here is deflected by the coarse titanium carbide particles that appear as protruding grains in the rough fracture surface. A theoretical prediction of composite toughening by a crack deflection mechanism has been proposed by Faber and Evans.[7] It indicates that crack deflection toughening is highly dependent on the dispersoid's

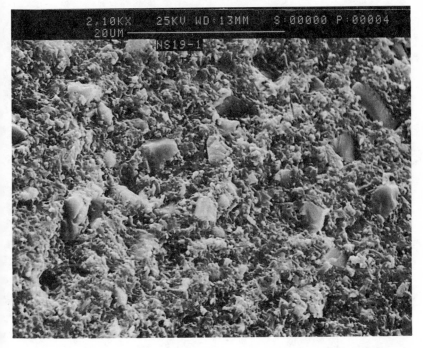

Fig. 1. Microstructure of a Si_3N_4/TiC composite with protruding TiC grains.

shape (Fig. 2). Thus a toughness increase of about 100% can be expected for particle composites, reinforced with spherical particles or small rods with an aspect ratio (length/diameter) lower than about 3. For composites reinforced by short fibres, such as whiskers that are characterised by an aspect ratio between 7 and 12, the toughness improvement is two times higher than with particles and can reach an overall improvement of four times the matrix toughness. Thus whisker composites are very attractive materials since they also exhibit other useful mechanical properties and can be manufactured with relatively simple processes, as will be shown later. The model of Faber and Evans[7] also indicates that disc-shaped particles should have a considerable effect on composite toughening, close to that predicted for short fibre reinforcement. For that reason, ceramics reinforced by hexagonal boron nitride particles have been developed[20, 21] for applications requiring thermal shock resistance.

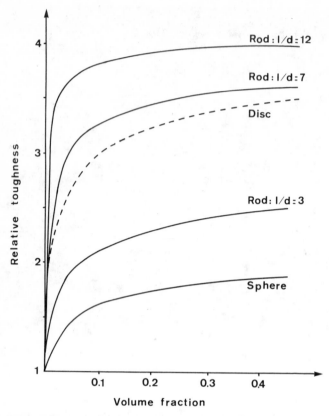

Fig. 2. Theoretical predictions of composite toughening by crack deflection.[20]

2.6 Fibre Pull-out

In the case of fibres or elongated particles with a high transverse toughness as the second phase, the stress field near the crack tip can induce debonding of the matrix–dispersoid interfaces leading to a pull-out mechanism.[2] An example of fibre pull-out is shown in Fig. 3 for silicon carbide whisker (SiC_w) reinforced alumina.

2.7 Crack Bridging

Finally, crack bridging is a very important toughening mechanism in elongated particle or short fibre (whisker) reinforced ceramics, that does not take place in the vicinity of the crack-tip.

In contrast to the other mechanisms, crack bridging affects a large

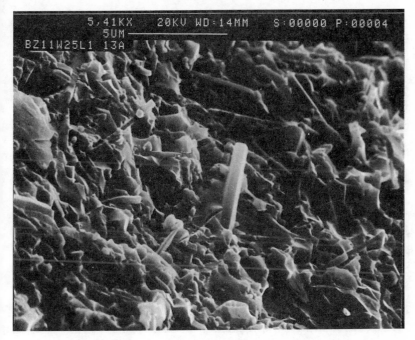

Fig. 3. Example of whisker pull-out in an Al_2O_3/SiC_w composite.

area behind the crack tip. A model of this mechanism has been developed using experiments on coarse-grained alumina.[21] The toughening results from the interlocking of the two fracture surfaces of the crack by the dispersoids. They induce closure forces on the crack surfaces, leading to a reduction in the stress intensity factor.[22,23] Bridging effects are often detected in whisker composites. The toughening of ceramics by whiskers has been analysed by Rühle and co-workers[24] who confirm that crack bridging is an important contribution in whisker toughening of a ceramic matrix. A typical microstructure of SiC_w reinforced alumina with crack bridging is shown in Fig. 4. Crack bridging has been observed to be independent of temperature and should act in the creep fracture region. Thus whisker composites have been reported to exhibit high creep resistance compared to monolithic ceramics.[25]

The different toughening mechanisms described above are often combined. The contribution of each mechanism is difficult to determine. Dominant mechanisms have been identified as dependent

Fig. 4. Microstructure of an Al_2O_3/SiC_w composite with crack bridging by SiC whiskers.

on the nature of the second phase, for example long fibres, particles, chopped fibres or whiskers. Some examples are reported in Table 1. It should be noted that an appropriate combination of second phases can lead to a greater improvement in mechanical properties since several major toughening mechanisms are involved. This is the case in alumina-based composites reinforced with zirconia particles and silicon carbide whiskers (Table 1).

The potential applications of this type of composite in various industrial fields have recently led to an increasing number of studies.[26-28] But ceramic matrix composites can also exhibit other enhanced properties. For instance, particle composites with refractory metal carbide and nitride dispersoids show concurrent improvement of both fracture toughness and hardness.[29] This is the reason for the development of titanium carbide reinforced alumina and silicon nitride based composites for cutting tool applications.[30] Another attractive property of ceramic composites containing refractory carbide or nitride dispersoids is the high electrical conductivity of these materials compared to the

TABLE 1

Comparison of toughening mechanisms for different ceramic composites (●, probable major factors of toughening; ○, other expected important mechanisms)

Second phase	Long fibres		Particles		Whiskers	Particles and whiskers
Example of composites	C/C	Glass/C	Al_2O_3/ZrO_2	Si_3N_4/TiC	Al_2O_3/SiC_w	$Al_2O_3/ZrO_2/SiC_w$
Expected toughening mechanism						
Load transfer	●	●				
Prestressing		○	○			○
Phase transformation			●	○	○	○
Microcracking	○		●		○	●
Crack deflection or	○	○	○	●	●	●
Crack impediment	○	○			●	●
Crack bridging	○				●	●

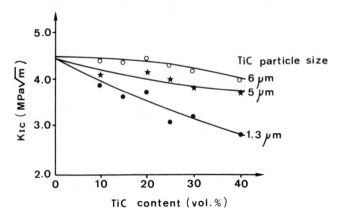

Fig. 5. Variations of Si₃N₄/TiC fracture toughness as a function of TiC content, for various TiC particle sizes.[29]

matrix. Conductive dispersoids, such as silicon carbide and titanium carbide/nitride particles can contribute to the improvement of the electrical conductivity of ceramic composites to such an extent (10^{-1} to $1\ \Omega^{-1}\ cm^{-1}$) that electrical discharge machining (EDM) can be used.[31]

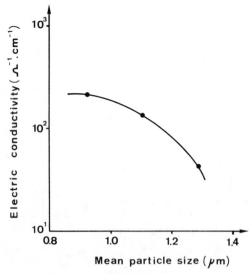

Fig. 6. Effect of TiN mean particle size on Si₃N₄/TiN (25 vol.%) electrical conductivity.[31]

However, the size of the conductive dispersoids must be carefully adjusted. It has been shown in the case of silicon nitride based composites that fracture toughness is a function of dispersoid size and that larger dispersoids yield greater toughening[29] (Fig. 5), while the electrical conductivity rapidly decreases as the conductive particle size increases[31] (Fig. 6). Various grades of electrically conductive ceramics are now being developed with good machinability by EDM.[31, 32]

3. MAJOR CERAMIC SYSTEMS

The ceramic matrix composites are an extended family of materials that can be classified with respect to the nature of the ceramic matrix, and to the nature or the shape of the second phase. The object of this section is not to give a complete compilation of all the ceramic matrix composites that have been reported in the literature, but to focus on ceramic composites of current interest that are the subject of industrial development or extensive research. A convenient classification of ceramic matrix composites consists of differentiating them according to the shape of the reinforcing phase. As indicated in Table 1, specific major toughening mechanisms may be associated with different types of second phase. It is thus convenient to differentiate long fibre composites from particle composites or whisker composites, which require different manufacturing processes, as will be discussed further.

High strength refractory fibres, such as carbon (C), silicon carbide (SiC), Al_2O_3 or Si_3N_4, are mostly used as a second phase for their physical and chemical stabilities and compatibilities at high temperature (generally higher than 1000 °C) during composite manufacture. Long fibres have primarily been incorporated in a glass or a glass–ceramic matrix, which exhibit the high elastic modulus mismatch that is required for an improvement by load transfer toughening.[34] The main characteristics of some long fibres have been described.[33, 34] Silicon carbide or silicon nitride whiskers are also commercially available and of current interest for the fabrication of whisker reinforced ceramics.[35] In the case of particle composites two types of particles are commonly incorporated in ceramic matrices, either refractory nitrides, carbides or borides, such as titanium carbide (TiC), silicon carbide (SiC) or boron nitride (BN), or zirconia (ZrO_2) particles for transformation toughening.

TABLE 2
Current status of the main existing ceramic matrix composites (●, industrial applications; ○, research work)

Second phase	Matrix					
	Al_2O_3	ZrO_2	Glass	Si_3N_4	SiC	C
Carbide, nitride or boride particles	●	○	○	●	○	
ZrO_2 particles	●	●	○	○		
Si_3N_4 whiskers				○	○	
SiC whiskers	●	○	○	○	○	
SiC fibres	○		●	○	○	
C fibres	○		●	○	○	●

The above reinforcing phases have been dispersed in various ceramic matrices, mainly Al_2O_3 as oxide matrix and Si_3N_4, or C, as non-oxide matrices. Glass or glass–ceramic matrices, such as lithium alumino-silicate, have also been used. Numerous compositions have been studied by combining these different matrices and dispersoids, but only a few composites are presently used for industrial purposes, or expected to be developed in the near future. A large part of the research and development work is focused on those materials that are listed in Table 2. One of the most important applications of ceramic matrix composites concerns cutting tools, and several composites are now commercially available for this purpose, such as alumina (Al_2O_3) based composites containing TiC and ZrO_2 particles[30] as well as SiC whiskers.[26, 36, 37] TiC reinforced silicon nitride (Si_3N_4) is also considered for such applications.[30] The aim of the development of Si_3N_4/TiC composites is to increase the hardness of the matrix while retaining the relatively high toughness which results from microstructural refinement.[1] Other potential applications are expected, especially for Al_2O_3/SiC$_w$ composites, in metallurgy, the engine industry (as rotary or stationary components of advanced gas turbine and diesel automotive engines[37] or valve-seats[8]) or in the aerospace industry.

Glass and glass–ceramic matrix composites are also under extensive study for aerospace applications and for development in the engine industry.[34] US Advanced Gas Turbine (AGT) research programmes consider SiC fibres or whisker reinforced glass matrix composites as good candidates for a range of applications, and a backplate made of glass–ceramic/SiC fibres has been tested.[28] Chopped SiC fibre reinforced silica has also been developed for portliner applications (Fig. 7) within the framework of an EUREKA European research programme (EU 29).[38]

Like/like composites, such as C/C or SiC/SiC, as well as SiC/C fibre composites, are also being considered for aerospace or engine applications. However, the high production cost of these materials does not lend itself to a large distribution market, such as that of the automotive industry. Various fields of application of C/C composites have been listed by Bracke *et al.*[69] An important development for C/C composites has been the brakes of aircraft and F1 racing cars.[39] For high temperature applications, the oxidation resistance of C/C composites is quite poor, and SiC/SiC and SiC/C fibre composites are more appropriate. Applications of such new materials will mainly concern the aerospace industry.[40]

The production cost of ceramic matrix composites is a very important criterion, since in many industrial applications they will be in competition with other conventional materials (metals, thermoplastics

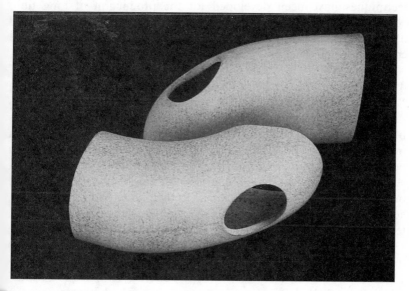

Fig. 7. Portliners made of SiC chopped fibre reinforced silica for automotive engines.

or monolithic ceramics) having low production costs. The automotive industry market is typical of this situation and the replacement of conventional materials by advanced ceramics or composites requires not only high performance and reliability but above all low production costs.

4. PROCESSING OF CERAMIC MATRIX COMPOSITES

The processing of ceramic matrix composites is more complex than that of monolithic ceramics because multiphase materials are concerned. Difficulties can be encountered in the forming process as well as in sintering. The reinforcing phase must be uniformly distributed in the ceramic matrix in order to avoid dispersoid clustering that would lower the second phase density in the surrounding region, thus giving greater matrix area for easy crack growth.[2] Difficulties in sintering can arise from mismatches in thermal expansion or in shrinkage between the second phase and the matrix,[41] so that hot-pressing techniques are often required for complete densification.

Various processing methods such as infiltration or physico-chemical techniques have been developed for the production of long fibre composites, while conventional ceramic processes are generally used after slight modification for particle, whisker and even chopped fibre composites.

4.1 Production of Long Fibre Composites

A frequent technique for the processing of long fibre composites is that of slurry infiltration[34,42] which consists of impregnating the fibres by passage through a slurry, prior to winding them to form a monolayered structure. The slurry is usually a mixture of a ceramic powder, a resin binder and water. Monolayer fibre tapes are then flat stacked to give multilayered materials with uni- or bi-directional fibre orientation. They are heated to burn out the binder and hot-pressed to achieve composite densification. For complete densification to occur, temperatures near or above the softening point of the matrix are required to allow viscous flow of the matrix between the fibres. This technique is more suitable for the reinforcement of glass or glass-ceramic matrices rather than of conventional ceramics.[43] Ceramic matrices imply higher processing temperatures which can damage the fibres or induce chemical reactions between fibres and matrix.[43]

—

However, in the case of reaction bonded silicon nitride (RBSN), an original process has been used for SiC fibre-reinforced RBSN, that consists of a silicon-based slurry infiltration followed by nitridation of the silicon particles at temperatures of approx. 1400 °C.[43, 46] The slurry infiltration leads to composites with densities generally higher than 98% of theoretical density.[34] Nevertheless, due to the number of operations and to the hot-pressing step it is a costly process[43] that can only be applied to components with simple geometry. Details of this technique have been reported by Prewo *et al.*[34] For the manufacture of complex-shaped components with the near-final shape and size, the melt infiltration technique is preferable. It consists of infiltrating a more or less complex shaped fibre preform by the molten matrix, using a hot-pressing procedure analogous to the squeeze casting techniques used for metal–matrix composites.[43] The matrix must be injected in the inter-fibre spaces at temperatures higher than the temperatures used for slurry infiltration, so that only glass matrices are of interest for this technique. Furthermore, due to the high viscosity of glasses compared to metals (about 1 kPa s at 1200 °C instead of about 1 mPa s at the same temperature for metals), slow infiltration occurs that leads often to excessive fibre–matrix chemical interaction.[34] Examples of glass matrix composite parts manufactured by this technique are reported in Ref. 34, with experimental details.

Chemical vapour infiltration (CVI) of fibrous materials is a relatively low temperature process for long fibre composite manufacture. CVI consists of flowing a reactant gas through a heated fibrous preform. The reactant gas decomposes and condenses around the fibres to form the ceramic matrix. Chemical reactions between fibres and matrix are avoided because of the low fabrication temperature. However, the deposition rates of the matrix are of the order of a few μm per hour or lower[47] and processing times of several hundred hours can be necessary for matrix deposition. Various modified CVI processes have been developed[44] in order to reduce processing times to ~24 h. Oxide and non-oxide matrices can be deposited by CVI, though most of the studies are focused on non-oxide matrices.[43] SiC/SiC, Si_3N_4/SiC and SiC/C composites can be prepared by this method, with a density that generally does not exceed 70–90% of the theoretical density.[29, 48] Thus they are characterised by only moderate oxidation resistance at high temperature and by limited mechanical strength,[29] but they exhibit a very high fracture toughness over a large temperature range.[48] Because of the long processing times, CVI remains a costly process that cannot be applied to the fabrication of ceramic composites for large scale markets.

Other chemical routes based on sol–gel techniques have been studied for the fabrication of long fibre composites.[43] Only laboratory scale processes are presently of concern, the industrial development of which will come up against various difficulties, above all economic difficulties because of the high cost of sol–gel components.

4.2 Production of Particulate or Whisker Composites

In the case of particulate or whisker composites, conventional ceramic processes can be used with minor modifications. The fabrication technique consists of an optimised mixing of the second phase, such as particles, whiskers or chopped fibres, into a ceramic powder slurry, followed by drying and sintering steps. The sintering usually requires the use of hot-pressing.

Difficulties can arise from the powder mixing since submicrometre matrix powders and reinforcing particles or fibres are generally used. The mixing of the two phases is the most important step of the process and can be conveniently achieved in a liquid medium. Organic or inorganic liquid medium may be used,[49] but for industrial purposes water is preferable. Low viscosity slips should be prepared with appropriate deflocculants and pH conditions.[49, 50] Mixing of the different phases can then be achieved and preference is given to ball mixing techniques.[51] Conventional turbomixing leads to inhomogeneous composites with agglomerates of one or other phase, such as those shown in Fig. 8 for Al_2O_3/ZrO_2 composites. They are characterised by poor mechanical properties (Table 3) compared to homogeneous composites obtained using ball milling (Fig. 9).

In the case of whisker reinforced composites the mixing procedure is more complex and consists of the dispersion of 50–200 μm sized agglomerates of closely entangled whiskers into submicrometre ceramic particles. Their mixing in liquid media requires the deflocculation of the different phases[51] with a combination of several techniques such as turbomixing, ultrasound and ball milling (Table 4). Conventional turbomixing is not sufficient to completely disperse the whisker agglomerates. The resulting composites are characterised by poor mechanical properties because of residual clusters of whiskers[52] (Fig. 10). Only the combination of turbomixing, ultrasonic and ball milling leads to optimised whiskers dispersion, as shown in Fig. 11, and to good mechanical properties (Table 4). However, these operations induce a significant reduction in whisker length[53] and must be carefully controlled.

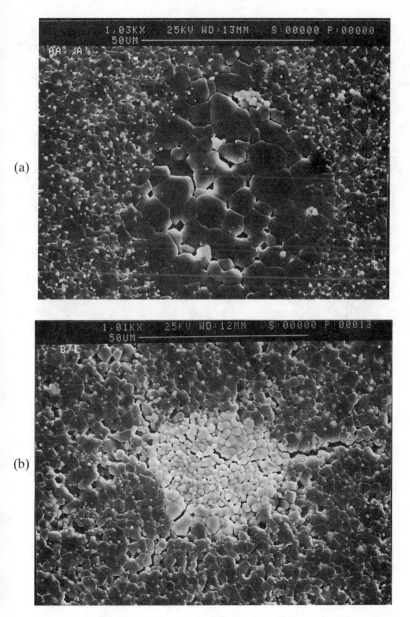

Fig. 8. Residual agglomerates in Al_2O_3/ZrO_2 composites. Porous alumina agglomerate with large grain growth (a). Zirconia agglomerate with micro-cracking (b).

B. Calès

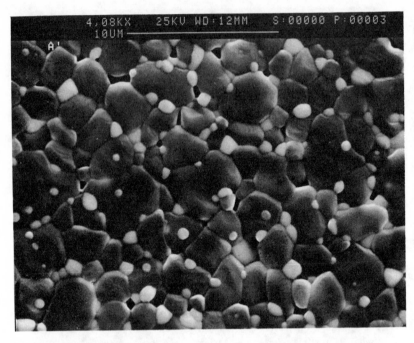

Fig. 9. Microstructure of an homogeneous Al_2O_3/ZrO_2 composite.

Prior to mixing the whiskers it is sometimes necessary to separate elongated whiskers from impurities, such as large SiC particles. Whiskers composites prepared using non-separated whiskers can exhibit poor mechanical properties due to large flaws formed by SiC particles[54] (Fig. 12). Appropriate sedimentation-flocculation techniques have been described for such separation.[54]

The sintering of particle composites can be achieved in some cases by

TABLE 3
Flexural strength of Al_2O_3/ZrO_2 (10 vol.%) prepared by using different mixing techniques

Mixing technique	Flexural strength 3 pt–20 °C
Turbomixing	340 MPa
Ball milling	630 MPa

TABLE 4
Influence of mixing techniques on mechanical properties of $Al_2O_3/ZrO_2 + SiC_w$ (20 vol.%)

Mixing technique			Flexural strength (MPa)		Fracture toughness $(MPa\ m^{1/2})$
Turbomixer	Ultrasonic	Ball milling	Average	Standard deviation	
+	−	−	762	100	4·6
+	−	+	604	50	5·3
+	+	−	755	36	6·7
+	+	+	880	55	7·0

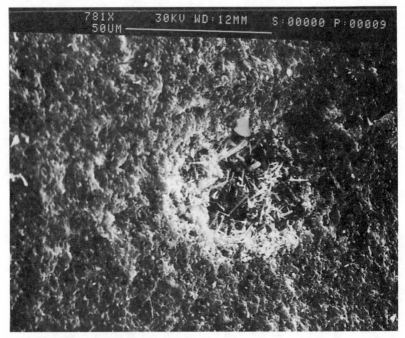

Fig. 10. Large void in Al_2O_3/ZrO_2 + SiC_w composite arising from a 50 μm sized whisker agglomerate.

using pressureless sintering at temperatures low enough to avoid chemical interaction between matrix and particles.

This is the case in ZrO_2 toughened Al_2O_3 and particle reinforced SiN_4.[40] Pressureless sintering with high heating rates, such as 400°C/min, has been used instead of hot-pressing to densify Al_2O_3/TiC composites.[55] Such heating rates can be conveniently achieved using microwave devices. Preliminary investigations in this field have demonstrated the possibility of particulate and whisker composite densification by microwave sintering.[56]

For whisker composites, uniaxial hot-pressing is usually used to reach high densification levels.[35] Pressureless sintering is only efficient for composites with low whiskers contents[57] and with short whiskers[58] which are characterised by moderate mechanical properties. Uniaxial hot-pressing of whisker composites induces an anisotropy in whisker orientation normal to the pressing axis and leads to composites with an orientation dependence of mechanical property.[59] Better values are

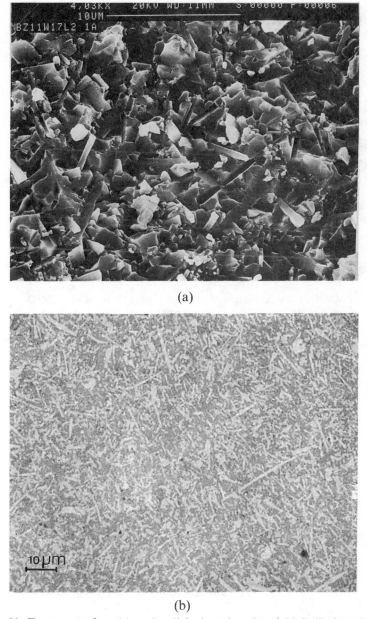

(a)

(b)

Fig. 11. Fracture surface (a) and polished section (b) of $Al_2O_3/ZrO_2 + SiC_w$ composite with a uniform whiskers distribution.

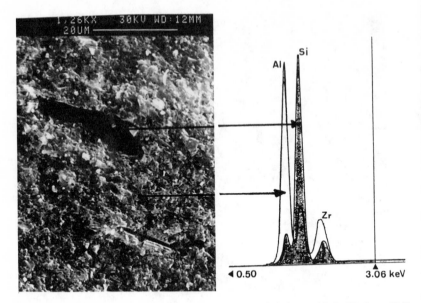

Fig. 12. Critical flaw consisting of a large SiC particle in an $Al_2O_3/ZrO_2 + SiC_w$ composite.

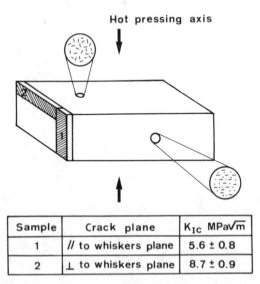

Sample	Crack plane	K_{Ic} MPa$\sqrt{m}$
1	// to whiskers plane	5.6 ± 0.8
2	⊥ to whiskers plane	8.7 ± 0.9

Fig. 13. Schematic of a hot-pressed whisker composite with the orientation of test bars and resulting fracture toughness obtained for Al_2O_3/SiC_w (20 vol.%).[59]

obtained when the crack plane is normal to the plane of the whiskers (Fig. 13). Consequently, the machining of whisker composite parts, such as cutting tools, from large hot-pressed blocks has to be adjusted to the whisker orientation.[60] In spite of these limitations, hot-pressing is commonly used for whisker composites manufactured to produce blocks as large as 120 mm in diameter and 30 mm in thickness.[52] Densification of whisker composites by hot isostatic pressing is under investigation[57, 61, 62] and will lead to more isotropic materials.

Microstructural analysis of hot-pressed SiC whisker composites reveals that there is no chemical interaction between the whiskers and the matrix.[53] Sometimes only a thin glass film is present along the whisker–matrix interfaces, that originates from the presence of an oxide-rich layer on the surface of the as-received SiC whiskers.[35, 53] For SiC whiskers with very low oxygen content, no glass phase has been detected at the whisker–matrix interface.[63] As a consequence, the chemical composition of SiC whiskers and especially their oxygen content, must be carefully controlled because it determines the nature and the thickness of the glassy fibre–matrix interface and can change the mechanical properties of the composite.

5. PROPERTIES AND ECONOMICS OF WHISKER COMPOSITES

Whisker composites are characterised by a combination of outstanding mechanical properties, as well as by good electrical and thermal properties. Thus, they are candidates for a number of applications in the field of cutting tools,[60] metallurgy, the automotive industry, and the aerospace industries. They are also candidates for high performance armour plate.

SiC whisker reinforced alumina, Al_2O_3/SiC_w has been the most extensively investigated, and two types of composites have been developed with either pure, or zirconia-toughened, alumina matrix. Both materials exhibit very good mechanical properties, but better properties have been reported for SiC whisker reinforced zirconia-toughened alumina[35] because of the multiple toughening produced by both zirconia particles (mainly transformation toughening) and SiC whiskers (mainly crack bridging, crack deflection and whisker pull-out). The variation of the fracture strength and toughness as a function of the content of reinforcing phase (zirconia particles plus SiC whiskers) are shown in Figs 14 and 15.[52] Flexural strengths higher than

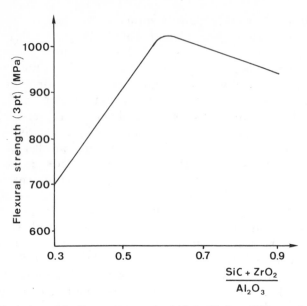

Fig. 14. Variation of the flexural strength of $Al_2O_3/ZrO_2 + SiC_w$ as a function of the dispersoid volume fraction.

1000 MPa were reached with a dispersoid content (ZrO_2 + SiC) of 38 vol.%. With higher dispersoid contents, the thermal mismatch between the second phase and the matrix leads to excessive residual stress in the composite, that decreases the fracture strength. On the other hand, the fracture toughness progressively increases with dispersoid content, and a value of 8 MPa $m^{1/2}$ has been obtained with a content of 44 vol.%. Higher toughness values have been reported by Claussen and Petzow[35] who have measured a toughness of 13·5 MPa $m^{1/2}$ in the composite $(Al_2O_3 + 15$ vol.% $ZrO_2) - 20$ vol.% SiC_w after annealing at high temperature. Such annealing modifies the characteristics of the fibre-matrix interface and leads to variations in mechanical properties. As previously mentioned, the characteristics of the fibre–matrix interfaces are controlled by the chemical composition of the SiC whiskers and especially by their oxygen content.

The influence of the SiC_w content on the fracture strength of Al_2O_3/SiC_w composites is shown in Fig. 16 for different whisker batches containing either 1·2 or 0·2 wt% of oxygen. A higher fracture strength is obtained for oxygen-free SiC whiskers. In that case, the Al_2O_3/SiC_w composites are characterised by whisker–matrix interfaces free of glassy

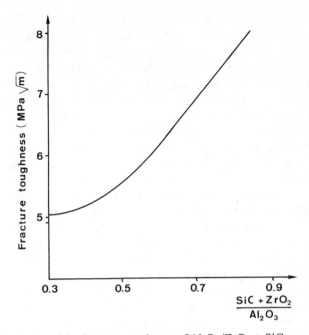

Fig. 15. Variation of the fracture toughness of $Al_2O_3/ZrO_2 + SiC_w$ as a function of the dispersoid volume fraction.

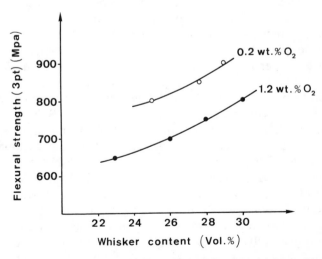

Fig. 16. Flexural strength of Al_2O_3/SiC_w composites for whiskers with different oxygen contents.

phase.[63] Moreover, as was first pointed out by Claussen and Petzow,[35] there is a synergic effect between transformation toughening due to zirconia particles, and whisker reinforcement leading to higher fracture strength and toughness for the multiphased $Al_2O_3/ZrO_2 + SiC_w$ composite, compared to binary composites, either Al_2O_3/SiC_w or Al_2O_3/ZrO_2.

The fracture strength of Al_2O_3/SiC_w composites remains unchanged up to $\sim 1000\,°C$ and the mechanical properties at room temperature are not modified after ageing at temperatures up to $1400\,°C$ for more than 100 h. Their thermal shock resistance is excellent and no decrease in fracture strength is observed for temperature changes up to $900\,°C$.[64] The improvement in thermal shock resistance of Al_2O_3/SiC_w composites compared to monolithic alumina is due to the interaction between the SiC whiskers and microcracks in the matrix, which prevents coalescence of the cracks into critical flaws.[64] The addition of whiskers also inhibits the high temperature creep of the alumina matrix, and a reduction of the creep rate at $1500\,°C$ of two orders of magnitude was measured for a 15 wt% SiC_w reinforced alumina composite.[25]

In addition to the above properties, SiC whiskers can also induce a significant increase of the thermal and electrical conductivities of the ceramic matrix that can facilitate the machinability of these composites by EDM techniques.

Whisker reinforced Si_3N_4 is also extensively described.[62, 65] However, the addition of whiskers is not as profitable for Si_3N_4 reinforcement as for Al_2O_3. The mechanical properties of monolithic Si_3N_4 ceramics are better than those of monolithic Al_2O_3 and the increased manufacturing cost involved in whisker addition is not justified by the improvement in fracture toughness and strength. The fracture strength of Si_3N_4 composites is generally decreased by SiC whisker additions[65, 66] and toughness improvement requires SiC whiskers with adjusted aspect ratio l/d to be really significant.[65] Before whisker reinforcement is attempted, microstructure refinement of monolithic Si_3N_4 leading to rod shaped grains should be thoroughly looked into, since it could generate a composite-like microstructure with high fracture toughness. Comparison of the microstructure of such a Si_3N_4, obtained by pressureless sintering, and with a toughness of $7·5$ MPa $m^{1/2}$ (Fig. 17), with the microstructure of an Al_2O_3/SiC_w composite (Fig. 11a), with the same toughness but obtained by hot-pressing, is striking. Thus, whisker reinforcement for Si_3N_4 ceramics improvement must be used only after considering carefully manufacturing costs and property improvements. The manufacturing cost of whisker composites is a decisive criterion in

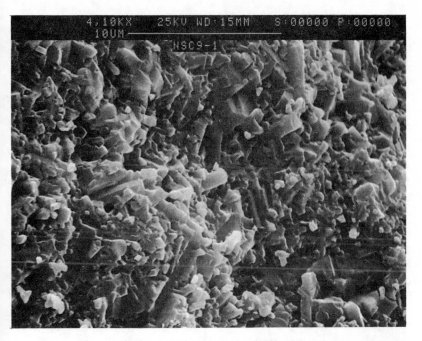

Fig. 17. Composite-like microstructure of Si_3N_4 ceramic with rod shaped grains.

future industrial developments. A detailed analysis of the economics of whisker composites has been performed by Karpman and Clarck.[67] They showed that the major cost drivers for whisker reinforced ceramics are sintering and machining costs (Fig. 18), while the cost of the mixing step which includes the cost of raw materials and especially whiskers is only 37% of the total cost. Therefore, improvements in the economics of whisker composites should be focused mainly on the sintering and machining steps, rather than on material cost reduction.[67] Concerning the sintering step, until the pressureless sintering is optimised, the development of hot isostatic pressing (HIP) processes for whisker composites manufacture will contribute to reduction of the production cost of this material. The use of EDM techniques could also reduce the cost of composite machining, and will allow the production of complex shapes. The cost of raw materials, mainly SiC whiskers, is directly related to the production volume of whiskers. The cost of \$18 kg^{-1}, used for the analysis of the production cost of Al_2O_3/SiC_w composites (Fig. 18), refers to a small production volume of around 2·5–3 t year^{-1}. Larger

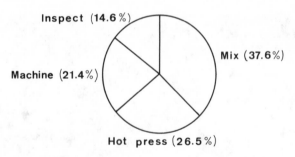

Fig. 18. Production cost of Al_2O_3/SiC_w cutting tool inserts assuming a production of 200 000 inserts and a cost of \$18 kg^{-1} for SiC whiskers.[67]

production volumes of whiskers are expected in the near future for the reinforcement of both ceramic and metallic matrices.[26, 68] They should probably be greater than 20 t year^{-1}, so that, according to the analysis of Karpman and Clarck,[67] the whisker cost should be reduced to \$4·5 kg^{-1} or less. Therefore, the cost of the mixing step for composite manufacturing (Fig. 18) will be appreciably decreased.

6. CONCLUDING REMARKS

Ceramic matrix composites are relatively new materials opening an expanding field of development. They are very promising for high temperature structural applications, while other applications at low or moderate temperatures require lower manufacturing costs of these materials. For this reason, ceramic matrix composites cannot compete for markets that can be filled by conventional monolithic ceramics. But specific markets that require exceptional properties, such as in metal cutting and forming, aerospace and military applications, could still provide a substantial field of development for ceramic matrix composites. Further developments of ceramic matrix composites will mainly be subject to the reduction of manufacturing costs, and of the costs of sintering and machining. The development of EDM grades is an attractive solution for the reduction of the machining cost and conductive ceramic matrix composites are now commercially available. In the case of whisker composites, the use of hot isostatic pressing instead of uniaxial hot-pressing will also contribute to a reduction of the production cost of these materials and will help their marketing. Microwave sintering is also considered for the pressureless sintering of ceramic composites. The reduction of production costs for long fibre composites is also a key factor for their industrial development. Improvement in the economics of long fibre composites will require modifications of forming processes, or new processes that can lead, through a reduced number of steps, to dense and near net-shaped composite parts.

REFERENCES

1. Tani, E. *et al*. Effects of size of grains with fibre-like structure of Si_3N_4 on fracture toughness, *J. Mater. Sci. Lett.*, **4** (1985) 1454–6.
2. Rice, R. W. Mechanisms of toughening in ceramic matrix composites, *Ceram. Engng Sci. Proc.*, **5** (1985) 589–607.
3. Ceramic composite issue, *Am. Ceram. Soc. Bull.*, *Special issue*, **65** (1986) 288–380.
4. Ceramic composite issue, *Am. Ceram. Soc. Bull.*, *Special issue*, **66** (1987) 303–76.
5. Ceramic matrix composites, *Composites*, *Special issue*, **18** (1987) 86–163.
6. Marshall, D. B. and Evans, A. G. Failure mechanisms in ceramic-fiber-ceramic matrix composites, *J. Am. Ceram. Soc.*, **68** (1985) 225–31.
7. Faber, K. T. and Evans, A. G. Crack deflection processes, *Acta Metall.*, **31** (1983) 565–84.
8. Guide to selecting engineered materials, *Advanced Materials and Processes, Special issue* (1987) 82–3.
9. Evans, A. G. and Faber, K. T. Crack growth resistance of microcracking brittle materials, *J. Am. Ceram. Soc.*, **67** (1984) 255–60.
10. Evans, A. G., Toughening mechanism in ZrO_2 alloys. In: *Science and technology of zirconia II*, Claussen, N., Rühle, M. and Heuer, A. H. (eds), Columbus, Ohio, American Ceramic Society, 1984, pp. 193–212.
11. Rice, R. W. and Freiman, S. W. Grain-size dependence of fracture energy in ceramics: II, Models for noncubic materials, *J. Am. Ceram. Soc.*, **64** (1981) 350–4.
12. Evans, A. G. and Heuer, A. H. Transformation toughening in ceramics: martensitic transformations in crack-tip stress field, *J. Am. Ceram. Soc.*, **63** (1980) 241–8.
13. Budiansky, B., Hutchinson, J. and Lambroupolos, J. Continuum theory of dilatent transformation toughening in ceramics, *Int. J. Solids Struct.*, **19** (1983) 337–55.
14. Lange, F. F., Transformation toughening, *J. Mater. Sci.*, **17** (1982) 225–55.
15. Reed, J. S. and Lejus, A. M. Affect of grinding and polishing on near-surface phase transformation in zirconia, *Mater. Res. Bull.*, **12** (1977) 949–54.
16. Heuer, A. H., Claussen, N., Kriven, W. M. and Rühle, M. Stability of tetragonal ZrO_2 particles in ceramic matrices, *J. Am. Ceram. Soc.*, **65** (1982) 642–50.
17. Wilfinger, K. and Cannon, W. R. Processing of transformation-toughened alumina, *Ceram. Proceed. ACS*, Sept./Oct. 1986, 13th Automotive Mat. Conf., Michigan, Nov. 1985.

18. Dutta, S. and Buzek, B. Microstructure strength and oxidation of a 10 wt% zittrite-Si_3N_4 ceramic, *J. Am. Ceram. Soc.,* **67** (1984) 89–92.

19. Lange, F. F. The interaction of a crack front with a second phase dispersion, *Phil. Mag.,* **22** (1970) 983–92.

20. Mazdiyasni, K. S. and Ruh, R., High/low modulus Si_3N_4–BN composite for improved electrical and thermal shock behaviour, *J. Am. Ceram. Soc.,* **64** (1981) 415–19.

21. Goeuriot-Launay, D., Brayet, G. and Thevelot, F. Boron nitride effect on the thermal shock resistance of an alumina-based ceramic composite, *J. Mater. Sci. Lett.,* **5** (1986) 940–2.

22. Swanson, P. L. *et al.*, Crack-interface grain bridging as a fracture resistance mechanism in ceramics, *J. Am. Ceram. Soc.,* **70** (1987) 279–94.

23. Marshall, D. B. and Ritter, J. E. Reliability of advanced structural ceramics and ceramic–matrix composites — a review, *Am. Ceram. Soc. Bull.,* **66** (1987) 309–17.

24. Rühle, M., Dalgleish, B. J. and Evans, A. G., On the toughening of ceramics by whiskers, *Scripta Metall.,* **21** (1987) 681–6.

25. Porter, J. R., Lange, F. F. and Chokshi, A. H. Processing and creep performance of SiC-whiskers-reinforced Al_2O_3, *Am. Ceram. Soc. Bull.,* **66** (1987) 343–7.

26. Key advanced ceramic markets — Part II, *High-Tech Materials Alert,* August (1987), pp. 5–7.

27. Multitoughening ceramic, *Technical Ceramics Bulletin,* **2**, 2 (1987) 14.

28. Advanced engine project of US Department Energy, *Industrie Céramique,* **815** (1987) 232–5.

29. Buljan, S. T. and Sarin, V. K. Silicon nitride-based composites, *Composites,* **18** (1987) 99–106.

30. Buljan, S. T. and Sarin, V. K. Silicon nitride-based composites. In: *Sintered metal ceramic composites,* Upadhyaya, G. S. (ed.), Amsterdam, Elsevier, 1984, pp. 455–68.

31. Kamijo, E. *et al.* Electrical discharge machinable Si_3N_4 ceramics, *Sumitomo Electric Technical Review,* **24** (1985) 183–90.

32. SiC-Si_3N_4 composite, *High-Tech Materials Alert,* August (1986) 9.

33. Mah, T., Mendiratta, M. G., Katz, A. P. and Mazdiyasni, K. S. Recent developments in fiber-reinforced high temperature ceramic composites, *Am. Ceram. Soc. Bull.,* **66** (1987) 304–17.

34. Prewo, K. M., Brennan, J. J. and Layden, G. K. Fiber reinforced glasses and glass-ceramics for high performance applications, *Am. Ceram. Soc. Bull.,* **65** (1986) 305–13.

35. Claussen, N. and Petzow, G. Whisker-reinforced oxide ceramics, *J. Physique,* **C1** (1986) 693–702.

36. Kolaska, H., Dreyer, K. and Reiter, N. *Property improvements in various ceramics through whisker reinforcement,* PM'86, Düsseldorf, July, 1986.

37. Double fracture toughness of ceramics, *Inside R & D,* June (1987) 3.

38. Rothmann, E. R. and Torre, J. P., *The use of ceramics in automotive engines,* Present Status and Development of Ceramics in Mechanical Industries, Saint-Ouen, France, June 1987.

39. Broquere, B., *From carbon–carbon composites to carbon–ceramic composites*, Ceramic–Ceramic Composites, Mons Belgium, April 1987.
40. Petiau, C. and Verneuil, J. C., *Thermal insulation of Hermes shuttle*, Thermal Transfer at High Temperature, Chatenay-Malabry, France, May 1987.
41. Brook, R. J., *Stress development during the sintering of composite ceramic systems*, Ceramic–Ceramic Composites, Mons, Belgium, April 1987.
42. Guo, J. *et al.* Carbon fibre-reinforced silicon nitride composite, *J. Mater. Sci.*, **17** (1982) 3611–16.
43. Cornie, J. A. *et al.* Processing of metal and ceramic matrix composites, *Am. Ceram. Soc. Bull.*, **65** (1986) 293–303.
44. Stinton, D. P., Caputo, A. J. and Lowden, R. A. Synthesis of fiber-reinforced SiC composites by chemical vapor infiltration, *Am. Ceram. Soc. Bull.*, **65** (1986) 347–50.
45. Coyle, T. W., Guyot, M. H. and Jamet, J. F., Mechanical behaviour of a microcracked ceramic composite, *Ceram. Engng Sci. Proc.*, **7** (1986) 947–57.
46. Corbin, N. D., Rossetti, G. A. and Hartline, S. D. Microstructure-property relationships for SiC filament-reinforced RBSN, *Ceram. Engng Sci. Proc.*, **7** (1986) 958–68.
47. Fitzer, E. and Gadow, R. Fiber-reinforced silicon carbide, *Am. Ceram. Soc. Bull.*, **65** (1986) 326–35.
48. Lamicq, P. J., Bernhart, G. A., Dauchier, M. M. and Mace, J. G. SiC/SiC composite ceramics, *Am. Ceram. Soc. Bull.*, **65** (1986) 236–8.
49. Wilfinger, K. and Cannon, W. R. Processing of transformation-toughened alumina, *Ceram. Engng Sci. Proc.*, **7** (1986) 1169–81.
50. Takas, F., Cannon, W. R. and Danforth, S. C. Colloidal processing of a SiC whisker-reaction bonded Si_3N_4 composite, *Ceram. Engng Sci. Proc.*, **7** (1986) 990–3.
51. Mathieu, P. and Calès, B., *Processing and properties of whiskers reinforced zirconia-toughened alumina*, Ceramic–Ceramic Composites, Mons, Belgium, April 1987.
52. Calès, B., Mathieu, P. and Torre, J. P., *Preparation and characterization of whiskers reinforced zirconia toughened alumina*, Science of Ceramics 14, Canterbury, Sept. 1987.
53. Sarin, V. K. and Rühle, M. Microstructural studies of ceramic–matrix composites, *Composites*, **18** (1987) 129–34.
54. Lunberg, R. *et al.*, Processing of whiskers-reinforced ceramics, *Composites*, **18** (1987) 125–7.
55. Borom, M. P. and Lee, M. Effect of heating rate on densification of alumina–titanium carbide composites, *Adv. Ceram. Mater.*, **1** (1986) 335–40.
56. Blake, R. D. and Meet, T. T. Microwave processed composite materials, *J. Mater. Sci. Lett.*, **5** (1986) 1097–8.
57. Tiegs, T. N. and Becher, P. F. Sintered Al_2O_3-SiC whisker composites, *Am. Ceram. Soc. Bull.*, **66** (1987) 339–42.
58. Hoffmann, M. J., Greil, P. and Petzow, G., *Pressureless sintering of SiC whisker reinforced silicon nitride*, Science of Ceramics 14, Canterbury, Sept. 1987.

59. Becher, P. F. and Wei, G. C. Toughening behaviour in SiC-whisker-reinforced alumina, *J. Am. Ceram. Soc.,* **67** (1984) C267–C269.
60. Vigneau, J. and Bordel, P. Influence of the microstructure of the composite ceramic tools on their performance when machining nickel alloys, *CIRP Annals,* **36** (1987) 13–16.
61. Lundberg, R. *et al., Glass encapsulated HIP-ing of SiC whisker reinforced ceramic composites,* International Conference on Hot Isostatic Pressing, Lülea, Sweden, June 1987.
62. Takemura, H., Miyamoto, Y. and Koizumi, M., *Fabrication of dense Si_3N_4-SiC whisker composite without additives by HIP-ing,* International Conference on Hot Isostatic Pressing, Lülea, Sweden, June 1987.
63. Sainfort, P., *Ceramic-ceramic composites,* Cegedur-Pechiney Internal Report, 1987.
64. Tiegs, T. N. and Becher, P. F. Thermal shock resistance of an alumina-SiC whisker composite, *J. Am. Ceram. Soc.,* **70** (1987) C109–C111.
65. Lundberg, R., Kahlman, L., Pompe, R. and Carlsson, R. SiC-whisker reinforced Si_3N_4 composites, *Am. Ceram. Soc. Bull.,* **66** (1987) 330–3.
66. Shalek, P. D., Petrovic, J. J., Hurley, G. F. and Gac, F. D. Hot-pressed SiC whisker/Si_3N_4 matrix composites, *Am. Ceram. Soc. Bull.,* **65** (1986) 351–6.
67. Karpman, M. and Clarck, J. Economics of whisker reinforced ceramics, *Composites,* **18** (1987) 121–4.
68. Ceramic composite licence opportunity, *High-Tech Materials Alert,* **4** (1987) 3.
69. Bracke, P., Schurmans, H. and Verhoest, J. Ceramic matrix composites. In: *Inorganic fibers and composite materials,* Oxford, Pergamon Press, 1984, pp. 97–121.

9

Ceramics in Internal Combustion Engines

Jürgen Huber and Jürgen Heinrich

Hoechst CeramTec AG, Selb, FRG

ABSTRACT

Current and future trends in structural automotive ceramics are discussed in this paper. The present state and the future of the adiabatic engine is scrutinized. The problem of the creation, and the benefits, of ceramic portliners are demonstrated with regard to turbocharger and exhaust gas cleaning catalyst performance. Ceramic components for the reduction of friction and wear are discussed. The application of wear resistant ceramics in the lubricant-free operating 'Ficht' engine is demonstrated. Information is supplied on current and future trends in making turbocharger rotors and on their future market prospects.

1. INTRODUCTION

New materials to be introduced into traditional markets ought to be at least partially superior to classical materials, a requirement to be met in order to overcome reluctance and scepticism on behalf of potential users. Ceramics are a group of materials which possess physical properties that could lead to considerable substitution for metals in certain areas of application. Some ceramic materials such as aluminium titanate, show low thermal conductivity and good thermal shock resistance, making them suitable for heat insulation purposes. Others, like silicon nitride are very hard, show good strength at high temperatures and also have good wear resistance. Many ceramic

Fig. 1. Fading device for a passenger car loudspeaker.

materials have a lower density than metals, thereby permitting some mass reduction. But despite their outstanding properties, oxide and non-oxide high performance ceramics have found up to now only limited application in internal combustion engines. According to a survey by Kamigaito et al.[1] about 85% of all ceramics used in Japanese cars are 'electro ceramics', being used for example as substrates and packages for microelectronic circuits and as piezoceramics. In Fig. 1 a fading device for passenger car loudspeakers with resistive and conductive paths printed on a ceramic substrate is shown. Such alumina substrates are important in constructing so-called hybrid circuits, being used for manufacturing compact board computers shown in Fig. 2. One source of information for the board computer is the fuel indicator supplying information on the fuel level, shown in Fig. 3. Such indicators are equipped with resistors printed on a ceramic substrate. The application of piezoelectrical materials, in anti-knock sensors for example, is shown in Fig. 4. These are important in automatic ignition readjustment systems. In safety belt tighteners, as

Fig. 2. Board computer constituted by hybrid circuits established on a ceramic substrate.

shown in Fig. 5, piezoceramics are used to improve passenger safety. Compared with 'electro-ceramics' the use of 'structural ceramics' is in an embryonic state in the automotive industry. Current and future developments in combustion engine ceramics are treated in this paper.

2. HEAT INSULATION CERAMICS

2.1 Motivation

A concrete example will illustrate the need for better heat insulation in automotive engines. In the heat balance of a Daimler-Benz M 110 E Otto engine operating at rated power shown in Fig. 6 it is seen that only 30% of the fuel energy can be used for direct car propulsion. The main

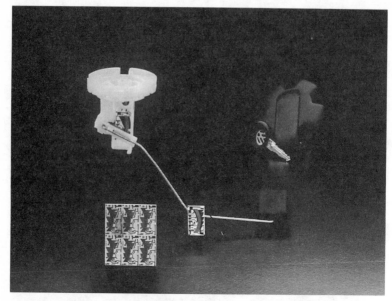

Fig. 3. Fuel indicator with resist printed on a ceramic substrate.

Fig. 4. Piezoceramic knocking sensor.

Fig. 5. Safety belt tightener.

energy losses are caused by heat transfer to the cooling media (15%) and to the exhaust gas (42%). If the engine is operating at a lower load, the effective power is even less. The losses by cooling and heat convection are increased, whereas the heat energy of the exhaust gas is almost halved. Measures to improve insulation could help to reduce energy

	Rated power	10%-rated power (pe = 2 bar at 2500 min^{-1})
Fuel energy	100% = 450 kW	100% = 62 kW
Cooling water	15% = 67·5 kW	28% = 17·5 kW
Lubricant	5% = 22·5 kW	8% = 5·5 kW
Exhaust gas	42% = 189 kW	22,5% = 3·5 kW
Mechanical losses +Auxiliary units +Heat emission +Heat convection	8% = 36 kW	25% = 15·5 kW
Effective energy	30% = 135 kW	16% = 10 kW

Fig. 6. Heat balance of a Daimler Benz M 110 E Otto engine, e.g. 6-cylinder Otto engine M 110 E Daimler-Benz.

losses caused by heat emission and convection, and by energy flow to the cooling media. Ceramic insulation is expected to lead to reduced fuel consumption rates, increased reliability, durability and system power density, lower service and maintenance effort, as well as decreased emission levels. Thermal insulation materials should have low thermal conductivity, a low thermal expansion coefficient, good themal shock resistance and good corrosion resistance, requirements that are met by ZrO_2 and especially by Al_2TiO_5.

2.2 The Adiabatic Engine

The most comprehensive insulation approach is to 'ceramize' the whole combustion chamber, producing the so-called 'adiabatic' engine. Expectations on the performance of adiabatic engines range at a very high level. In 1985 the Japanese automobile manufacturer Isuzu Motors Ltd announced that a ceramic turbocompound engine could have a 30%

Fig. 7. Computer picture of the Porsche engine, showing the position of the portliner.

increase in efficiency, compared to a conventional diesel engine. But computer model calculations for turbocompound engines with ceramic combustion chambers have shown that, in the case of passenger car diesel engines with a small displacement, only marginal improvements can be achieved in efficiency and fuel consumption.[2] The Ford Motor Co. has claimed a possible 60% reduction in fuel consumption by using a ceramized adiabatic diesel engine, properly designed for the use of ceramics. But these and other optimistic expectations can not yet be verified experimentally. In contrast, several research groups have shown that ceramic insulated combustion chambers produce disadvantageous effects. Theoretical calculations suggest that increased combustion chamber temperatures would lead to an improved degree of efficiency. Experimental investigations by Woschni *et al.* have shown that wall temperatures above 700 °C increase the heat flow from the working gas to the wall. Thus energy is lost leading to decreased efficiency and increased fuel consumption.[3] Another group also obtained a decrease in efficiency by using combustion chambers with an 80% coating of aluminium titanate and ZrO_2.[4] The only positive effect is a slight reduction of the cooling effort.[5] These experimental results are somewhat contradicted by more encouraging practical experience with adiabatic diesel engine prototypes used in heavy duty engines. Therefore the future of adiabatic engines remains rather uncertain, but is still a focus of research activity.

2.3 The Portliner

Another approach to the use of low thermal conductivity ceramic materials is the insulation of certain engine compartments, where considerable heat losses are to be expected. Such a position is the cylinder head exhaust channel. Lining of this channel with ceramic thermal insulation materials should lead to technical and economical improvements in engine performance. The position of those units, called portliners, can be seen in Fig. 7, which is a computer picture of the Porsche engine. The development, design and production of portliners has been the subject of close cooperation between Hoechst CeramTec AG and Porsche AG. Since 1985 portliners produced by Hoechst CeramTec AG have been introduced into the serial production of the Porsche 944 Turbo engine. A review of the development history of the portliner supplies an example of the problems and difficulties that designers, ceramic producers and engine manufacturers have to meet.

The first problem to be solved in the creation of the portliner was the

		Al$_2$TiO$_5$	Zirconia	Cast Iron	Aluminium
Density	(g/cm^3)	3·15	5·8	7·25	2·7
Young's modulus	(GPa)	15	200	120	70
Heat conductivity	(W/mK)	2	2·5	58	220
Coefficient					
of expansion	(1/K ·10^6)	1	10	10·5	23·8
Bending strength	(MPa)	30	500	250–600	100–600[a]

[a]Tensile strength.

Fig. 8. Physical properties of Al$_2$TiO$_5$, ZrO$_2$, aluminium and grey cast iron.

development and optimization of ceramic materials appropriate to the functional requirements. In Fig. 8 the essential physical properties of aluminium titanate (ATI) and zirconia are listed. To manufacture cylinder heads with ceramic portliners, aluminium or cast iron has to be cast around the ceramic, and the properties of those materials are also shown as reference data. Firstly ATI possesses a low density, which is in the range of aluminium and much lower than that of ZrO$_2$, thus favouring ATI against zirconia in terms of mass reduction. ATI also shows a very low thermal expansion coefficient, which is about ten times lower than the corresponding value for zirconia. This low thermal expansion coefficient is responsible for the good thermal shock resistance of ATI. The low measured, overall expansion coefficient of polycrystalline ATI is caused by an anisotropy of the crystal structure with a strong negative coefficient in one crystallographic direction. In the ceramic this anistropy also leads to considerable structural stresses during cooling, thus causing cracks within the structure for tensile stresses exceeding a critical level. If an ATI body is heated, the corresponding dilatation of the material is absorbed by the micro-crack system, resulting in a very low expansion over a certain temperature range. These microcrack phenomena are also responsible for the very low Young's modulus, which gives a pseudo-elastic character to ATI. Due to this fact, metals can be cast around ATI parts, whereas zirconia parts cannot sustain the stresses caused by the solidification of the metal. A disadvantageous property of ATI is its low bend and tensile strength, which is reduced by more than a factor of 10 compared with other ceramic materials. Because of this low strength, caused by structural microcracks, ATI cannot be used as a structural material on its own. It has to be combined with other materials guaranteeing a better

Wärmezufluß / heat flux   Isolation / insulation

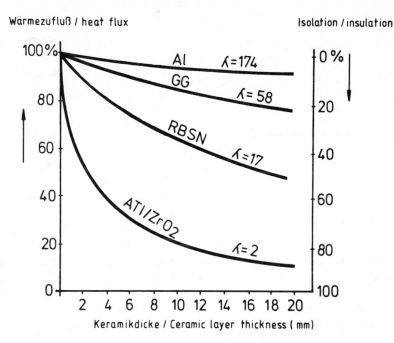

Fig. 9. Heat flux diagram of Al_2TiO_5, ZrO_2, Si_3N_4 and grey cast iron.

mechanical strength of the whole part. Most important for materials being used in ceramic portliners are their thermal conduction properties. ATI and zirconia possess very low conductivities, ranging between 2 and 2·5 $Wm^{-1} K^{-1}$, thus 20–30 times lower than cast iron and about 100 times lower than aluminium. Figure 9 shows the heat flux through a ceramic layer or its corresponding insulation as a function of the wall thickness. Its essential point is that low thermal conductivity decreases the wall thickness needed to obtain a specified degree of insulation. Considering all the material properties discussed, ATI was chosen as a construction material in developing the ceramic portliner.

The original design of the Porsche 944 Turbo engine did not include a portliner unit. Therefore it was necessary to adapt the portliner design to the cylinder head geometry of the original design, the valve size and the requirements being imposed by fluid mechanics. The design meeting all those requirements is shown in Fig. 10. As mentioned above ATI cannot be used as a structural material on its own. It has to be used in conjunction with a supporting material. This can be achieved by casting

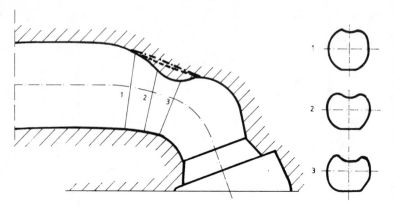

Fig. 10. Section of the exhaust valve of the Porsche 944 turbo engine.

round it the aluminium or cast iron. Casting round a ceramic portliner designed according to Fig. 10 would introduce considerable bend and tensile stresses, caused by shrinkage during solidification of the metal, and leading to destruction of the element. Therefore it was necessary to create a portliner geometry which was only exposed to compressive

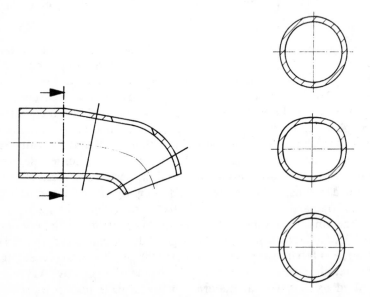

Fig. 11. Section of portliner design adjusted to ceramic material properties.

stresses. Such a design, not completely optimized for fluid-mechanical requirements and cylinder head geometry, can be seen in Fig. 11. The key feature of this design is the concave shape of all curves. Portliners designed according to Fig. 11 were well-suited to the casting procedure. This portliner design was only marginally altered in some detail for mass production for the Porsche 944 engine.

Before reaching serial production level the portliners had to survive several tests to prove their good reliability. A lot of other problems had to be solved. For example, very narrow tolerances had to be met in production. As with many other multi-step processes, alterations in a variety of parameters result in dimensional deviations. Almost all ceramic materials are subject to shrinkage during sintering. Thus some sophistication of the sintering process was necessary to reach appropriate shrinkage. Raw materials also play a key role in the accurate production of components. Strict and specified control of the starting materials is therefore a very important measure. Most production steps had to be extensively automated.

What are the technical benefits of ceramic portliners? To answer this question it is necessary to know something about the influence of portliners on the engine energy flow. First of all, there will be no change in fuel consumption and effective power, as combustion processes are not altered by portlining. Convection and emission losses will be influenced only in a marginal way. But there is a certain kind of redistribution of energy between cooling and exhaust gas. In the Porsche 944 engine ceramic portliners of about 3 mm thickness reduce the heat flow from the exhaust port to the cooling by up to 7 kW, thus causing a 13% decrease of required cooling effort, thereby permitting some mass reduction in the total cooling system.[6] The exhaust gas energy flow is enhanced by 8 kW at full load, leading to an exhaust gas energy increase of about 3%[6] compared to an engine without portliners. This is especially favourable in an engine equipped with a turbocharger unit. In Fig. 12 comparing the exhaust gas temperature of the Porsche 944 Turbo engine with and without portliner one can see that the exhaust gas is hotter by more than 30°C at full load.

Increased exhaust gas temperatures are also important for exhaust gas catalysts. The activity of the catalyst is temperature dependent. Only after reaching a certain temperature threshold is the catalyst working with 100% efficiency. Reaching this temperature as quickly as possible is an essential feature in reducing harmful gas levels, especially during the cold starting period. Due to the use of the portliner the hydrocarbon

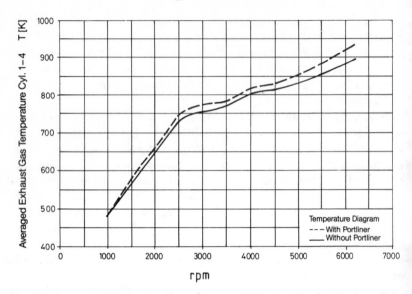

Fig. 12. Averaged exhaust gas temperature of the Porsche 944 Turbo engine
(with and without portliner).

level in the exhaust gas has been reduced by 16%, and the CO and NO_x
contents are lowered by about 10%. As diesel engines also have low
exhaust gas temperatures ceramic portliners could find some application
in diesel passenger cars, heavy-duty engines and tractors, especially
when the engines are equipped with turbochargers.

The beneficial results of portliners have to be considered in
conjunction with higher costs per cylinder head. The introduction into
serial production can be enhanced if mass production feasibility
guarantees high piece numbers. But a thorough problem oriented use/
cost analysis is always necessary.

2.4 Further Applications of Insulating Ceramics

Another interesting aspect is the shielding of the piston with ceramic
inserts or coating. Different materials such as ATI, ZrO_2, Si_3N_4 and even
SiC are used in the test procedure. In the literature[1,4,7,8] quite different
results are reported. Up to now it has not been decided to which extent
these measures are advantageous for engine performance.

Two different combustion processes apply in diesel engines, to be
distinguished by the way in which an intimate ignitable air/fuel mixture

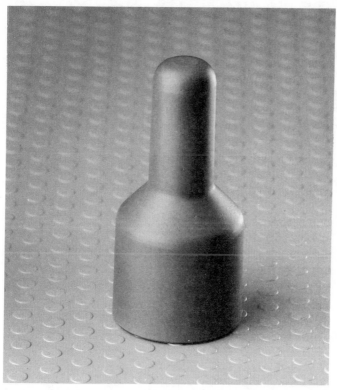

Fig. 13. Experimental precombustion chamber made of **HIPRBSN**.

; produced. In diesel engines equipped with swirl chambers the fuel is eated and evaporated before ignition. The combustion proceeds from he chamber, causing a maximum temperature at the periphery of the hamber. Swirl chambers made from superalloys are not durable nough to be used in higher performance engines. Therefore they have • be substituted by ceramic chambers. At least three Japanese utomotive manufacturers are commercializing ceramic swirl chambers.[1] Iazda Motors Ltd and Isuzu Motors Ltd are using them in diesel assenger cars and the application of all-ceramic inner wall swirl hambers by Toyota Motor Corp. was believed to have started in late •86.[1] Ceramic swirl chambers generate a higher boost pressure for the rbocharger unit, increase the power output and decrease the torque id the idle noise. As the emissions are cleaner, it may perhaps be

Test procedure

Property	4-point bending, 30 mm outer span 10 mm inner span samples diamond ground	c-ring test, samples cut from precombustion chambers, surface as fired	c-ring test, samples cut from precombustion chambers, oxidized 1200°C/60 hrs	c-ring test, samples cut from valve guides, surface as fired
Fracture strength, MPa	830	660	650	730
Weibull modulus	10	10	9	9
Young's modulus, GPa	310	292	290	266
Density, g/cm³	3·26	3·26	3·26	3·26
% of theoretical density	99·7	99·7	99·7	99·7
β-content, %	100	100	100	100

Fig. 14. Material properties of c-ring test samples cut from HIPRBSN.

possible to avoid the use of electronic fuel injection systems and diesel particulate traps, especially during the cold starting period.

In the precombustion process the fuel is sprayed onto a flame spreader thus causing a finer dispersion of the fuel. During combustion the temperature rises up to 1000 °C, a working condition which is manageable only with expensive metallic or ceramic materials having good high temperature strength and thermal shock resistance. Figure 13 shows an experimental precombustion chamber produced for Daimler-Benz AG made of hipped reaction bonded silicon nitride (HIPRBSN). The strength of metallic super alloys diminishes in the temperature range above 800 °C, whereas HIPRBSN still shows sufficient strength at 1000 °C. Thus HIPRBSN is a promising material for ceramic pre-combustion chambers.

In Fig. 14 the material properties for the 4-point bend test and c-ring test samples cut from HIPRBSN precombustion chambers and valve guides are shown. In one case the samples were cut after sintering and in the other the chamber was exposed to oxidizing conditions at 1200 °C for 60 h. There are only slight changes in material properties proving the good durability and resistance of HIPRBSN.

3. CERAMIC COMPONENTS FOR REDUCTION OF FRICTION AND WEAR

Other advantageous properties of silicon nitride are its good wear resistance, due to its high hardness and its low density. Other materials do not have such a broad spectrum of favourable properties. Therefore, HIPRBSN is an especially promising material for mass reduction in wear and friction exposed, as well as oscillating, parts of the engine.

3.1 Ceramics in the Valve Mechanism

Valve guides are among the most wear exposed combustion engine components, requiring good high temperature strength, wear resistance and corrosion, as well as good sliding properties. Valve guides made from partially stabilized zirconia (PSZ) were the first objects to be investigated for this purpose. Together with chromium plated steel valves they showed low wear rates.[9] PSZ valve guides have been shrink fitted into a grey cast iron housing. The serial use is not yet known. Nowadays valve guides made of Si_3N_4 are in an experimental stage.

Fig. 15. Valve guide made of HIPRBSN.

Figure 15 shows a valve guide made from HIPRBSN.

Figure 16 shows, besides the previously mentioned parts such as portliners, valve guides and a shielded piston, other parts which may be made from ceramics, like valves, valve spring retainers, valve seat rings, cylinder liner segments and piston bolts. The application of ceramic valves is, due to the possible mass reduction, a point of special interest. By using ceramic valves it will be possible to install valve springs which will be softened by 30%. This enables the application of a crank shaft equipped with more narrow tappets, allowing a considerable mass reduction of the crank shaft, thus reducing friction losses and cooling effort. Through all these measures the revolution numbers of the engine can be increased, for example, from 6000 rpm to about 7100 rpm leading to increased efficiency of the engine.

Valve spring retainers are used to transmit the spring force to the valves. Several companies are dealing with the ceramization of these

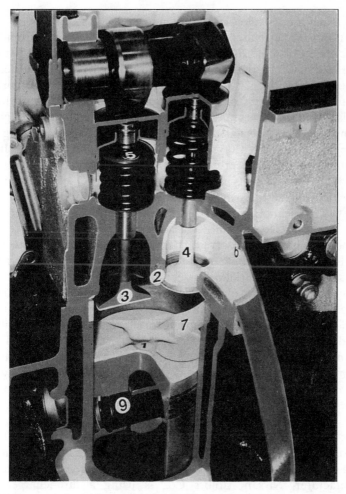

Fig. 16. Section of a Daimler engine demonstrating possible applications of structural ceramics. 1, ceramic shielded piston; 2, valve seat ring; 3, ceramized valve; 4, ceramic valve; 5, valve spring retainer, 6, portliner; 7, cylinder liner segment; 9, piston bolt.

parts in order to gain higher durability and mass reduction. Silicon nitride ceramics, HIPRBSN or SSN, are promising materials for these purposes, showing encouraging results in proof tests and in engine tests. The development of ceramic tappet inserts and segments of the cylinder liner is still in the experimental stage.

3.2 Ceramics in Non-lubricated Engines

Ceramics are also used in lubricant free engines, for example a two-stroke engine of about 20 kW developed by Ficht GmbH in cooperation with Hoechst CeramTec AG, shown in Fig. 17. In this engine lubrication of the piston with oil is not necessary. The lubricating systems for pistons and the altered crank drive are separated. The lower cylinder area is sealed from the crank case and the piston rod is led through the sealing, thus being insulated from the crank area. Within this engine a number of components have been made from ceramic materials. It was possible to use silicon carbide (SiSiC) ceramics for complete pistons, cylinder liners, and parts in the crank drive region. General advantages of the Ficht engine are environmental protection by avoiding combustion chamber lubrication with oil–fuel mixtures and a much lower noise level. Almost no corrosion of pistons, cylinders and other ceramic components has been observed. The power/mass ratio of this engine is very high and it requires a smaller volume compared to engines of similar type. Reduced friction losses and a considerable mass reduction through the application of ceramic materials results in an improvement of the degree of efficiency, thus reducing fuel consumption rates. The use of ceramic materials also contributes to a much improved reliability and durability of the engine. Further advantages are its good cold starting behaviour and lower production costs by reducing the number of components required, compared to conventional two-stroke engines.

4. THE TURBOCHARGER ROTOR

Turbochargers are used for gaining additional power in a short time by using exhaust gas energy. Up to now only one Japanese passenger car manufacturer, Nissan Motors Co., is using a turbocharger unit equipped with a ceramic rotor. This ceramic rotor made from silicon nitride and installed in the Fairlady Z model is still a compromise between durability and cost. Considering driving cycles common and typical for Japanese traffic, the lifetime and durability of those rotors is good enough. But especially in Europe the requirements for reliability and durability of ceramic rotors are much higher, because of differences in driving behaviour between Japanese and European drivers. Therefore it is necessary to make advances in material development, and in design.

For turbocharged engines an improvement of the response behaviour at the stationary state is essential. This could be achieved by

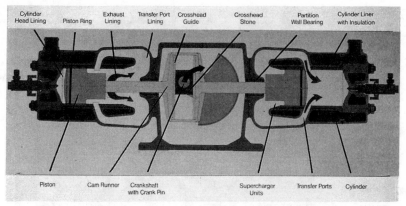

Fig. 17. Cross-section through the Ficht engine.

considerably reducing the mass and therefore the moment of inertia of the turbocharger rotor by using ceramic materials for its construction. Figure 18 shows the increase of the turbine revolution number versus time. Comparing metallic and ceramic rotors, ceramic rotors reach certain numbers of revolutions earlier than metallic ones. About 60% of the nominal speed is obtained in half the time using ceramic rotors. As mentioned before the exhaust gas temperatures of engines equipped

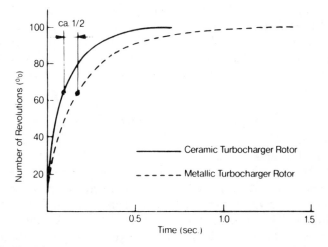

Fig. 18. Transient response behaviour of metallic and ceramic turbocharger rotors.

Fig. 19. Experimental turbocharger rotor made from HIPRBSN.

with insulating components made from ceramics are higher. The strength of superalloys used nowadays in turbocharger units is decreasing rapidly, reaching a temperature limit around 1000°C, a temperature at which Si_3N_4-materials especially still show satisfactory strength. Therefore ceramic turbocharger rotors are promising for application in engines with higher exhaust gas temperatures. An experimental turbocharger rotor made from HIPRBSN and developed by Hoechst CeramTec AG is shown in Fig. 19.

5. FUTURE MARKET PROSPECTS

5.1 General Remarks

There are several strongly varying assessments on future cost, demand and market developments for high performance ceramics with different results depending on the economic environment in which the forecasting model is situated. By the year 2000 the market for ceramic heat engines could be 1 billion dollars a year (billion = 10^9), at an annual growth rate of 40% per year.[10] For the same period of time a Charles Rivers Associates forecast predicts a 21·5% penetration rate into US market for cars equipped with ceramic components. Bowen (MIT)

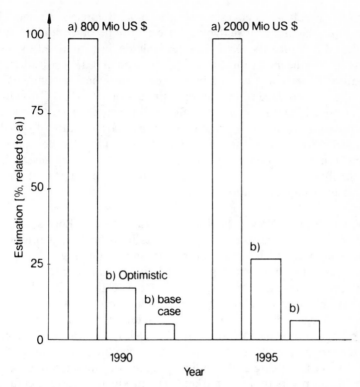

Fig. 20. Projections on the Japanese market for structural fine ceramics used in the engine region. (a) S. Saito, Japan Fine Ceramics Association; (b) other sources.

forecasts that at the end of this century 23 kg of ceramic parts will be installed on board a passenger car, thus constituting a 15 billion dollars-a-year-business.[10] Those pretty optimistic predictions are contradicted by less optimistic estimates. In Fig. 20 one can see a comparison between several estimates on the Japanese structural fine ceramics market for engine use. The study of the Japanese Fine Ceramics Association in which values are equalized to 100% in this index diagram expects the structural ceramics market volume for Japan to rise from 800 million dollars in 1990 to 2·0 billion dollars by the year 1995. The optimistic evaluation of a more conservative forecast predicts only about 20–25% of the market volume expected by the JFCA projection, whereas a base case study predicts a market volume representing only 4–5% of the figures forecast by JFCA. Therefore it is very difficult to interpret correctly and to rely on such general forecasting models, because of a lack of knowledge of those economic preconditions and technical developments heavily influencing the data obtained by such forecasts. Even the most thorough general projection bears a considerable amount of uncertainty, caused by economic parameters, competition with other materials, and changes in application demands. To reduce uncertainty and to diminish the number of influencing parameters it is more useful to focus on the prediction of the market future of specific ceramic parts, and to use them as a kind of a model for all other ceramic combustion engine components. Therefore the following discussion deals solely with the future cost, demand and marketing projections of ceramic turbocharger rotors.

5.2 Future Market Prospects of Ceramic Turbocharger Rotors

The perceived benefits of using ceramics in turbocharger rotors are mass reduction, high temperature strength, and high operating temperatures. Furthermore it is expected that ceramic rotors could cost less if certain requirements are met.[11] Despite the potential of the demonstrable, technical benefits, ceramics have only slowly penetrated the turbocharger market. Only one manufacturer is currently producing and selling ceramic turbocharger rotors on a large commercial scale. A research group at the MIT has worked out a detailed study, focussing on the influences fostering and handicapping the introduction of ceramic turbocharger rotors into the market.[11] The production cost for ceramic turbocharger rotors is based on injection moulding as the forming process, requiring expensive tools, and on sintering followed by hipping for enhancement of certain ceramic properties. The extremely long binder removal cycle times, and the finishing process, are also major features increasing production costs. The production costs are estimated for a rotor of a final mass of 60 g and a final specific density of $3 \cdot 2 \, Mg \, m^{-3}$. Figure 21 gives information about the interdependence between turbocharger rotor cost and overall turbocharger yields. At present, yields for ceramic turbocharger rotors may be as low as 10% overall. An increase in the yield from about 35% with approximately $140 per piece, to 70% reduces costs to a tenth, which would mean about $14 per piece. The costs are not only influenced by the production steps, injection moulding and sintering, but more importantly by binder removal, sintering and hipping, finishing and inspection. Improving yields and reducing costs in the final processing steps is important because value is added to the material throughout the production sequence, giving higher impact to yield failures during final processing.

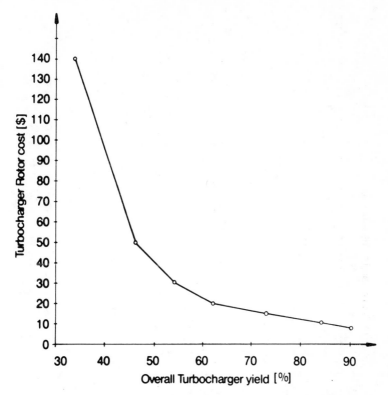

Fig. 21. Dependence of turbocharger rotor cost on overall yields. (Rothman *et al.*[11]

Furthermore it is necessary to increase production figures. This goal can be achieved by automating, standardizing and sophistication of most production steps, decreasing personnel costs and improving yields by avoiding individual mistakes. Another important key feature for increasing production levels is the existence of nondestructive evaluation methods working at a reasonable cost level and at a time scale adjusted to production cycle requirements. There is also an interdependence between production numbers and yields. For example, at a production volume of less than 10 000 pieces per year the yield limit is about 50%, whereas at a volume of more than 20 000 pieces/year a yield of 90% has to be achieved. Ceramic rotor costs become competitive above 70% overall yield which in this scenario represents a production volume of 100 000 rotors/year. At the present level of ceramics technology, ceramic

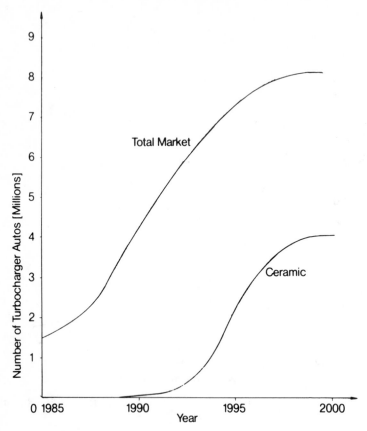

Fig. 22. Projected ceramics penetration into the turbocharger market. (Rothman et al.[11])

rotors cannot be manufactured with the required reliability and at a reasonable cost level. Another problem is the development of appropriate methods of joining between the ceramic rotor and the metal shaft. But besides costs, rotor performance, durability and reliability are key features in improving the situation of ceramic turbocharger rotors in competition with metallic ones.

The performance required of a turbocharger is dependent on the type of driving which differs between the US, Europe and Japan. Only luxury vehicles will be able to hide the additional costs caused by a ceramic rotor, and only improvements in technology could open the gate for the introduction of ceramic rotors to a wide range of vehicles. The automotive diesel engine requires power boosting for improved engine performance. Therefore this type of engine constitutes a promising market for the turbocharger. Additionally, due to its reduced mass, the engine lag could be reduced by ceramic rotors. There is also a large profit margin on turbocharger components for diesel engines, so some flexibility on the cost of the rotor could be tolerated for increased performance.

Although some turbocharger manufacturers have announced plans to use ceramic rotors, it is not very probable that they will be manufactured on a large scale before 1990. Currently, large-scale production facilities working with high yields presumably do not exist. Figure 22 shows the projected ceramics penetration into the turbocharger market. In the beginning, the market for ceramic rotors will be limited, ranging from about 10 000 rotors/year. Between 1990 and 1995 the penetration of ceramic rotors is expected to rise markedly reaching a 50% market share by the year 2000, because they are most desirable only in high-priced high-performance vehicles. Some contradictory market projections indicate a market size for ceramic turbochargers of 200 million $ in the year 1995,[12] which now seems very unlikely. Generally, an increasing total market for turbocharger equipped passenger cars is boosting ceramic rotor market penetration. The dilemma hampering expanded use of ceramic turbocharger units is that automotive manufacturers will refrain from using ceramic rotors on a large scale until they can be assured of reproducibility, excellent reliability, performance and durability at a reasonable cost level. A clear demonstration of the performance of silicon nitride in turbocharger operation is required to overcome those barriers.

6. CONCLUDING REMARKS

Even if there are many doubts as to the amount by which ceramics will penetrate the engine, it is likely that essential parts of the engine of the future will be made out of ceramics. But improvement in reliability, proper ceramic design, and the reduction of cost are necessary prerequisites. There are also strong development efforts by many car manufacturers to construct and build gas turbines. In the way projected, they can only be realized with ceramic parts.

REFERENCES

1. Kamigaito, O., Ceramic components for passenger car engines. In: *Materials science monographs: high tech ceramics*, Vincenzini, P. (ed.), Amsterdam, Oxford, New York, Tokyo, Elsevier, 1987, pp. 2489–502.
2. Klarhoefer, C., Rechnerische Untersuchungen von Compound-Anordnungen bei einem wärmegedämmten PKW-Dieselmotor. In: *Einsatzchancen keramischer Werkstoffe im Motorenbau*, Conference paper, Haus der Technik e.V. (ed.), 1987.
3. Woschni, G., Einfluß von Isoliermaßnahmen auf die Prozeßgrößen bei Dieselmotoren. In: *Einsatzchancen keramischer Werkstoffe im Motorenbau*, Conference paper, Haus der Technik e.v. (ed.), 1987.
4. Heinrich, H., Langer, M., Siebels, J. E., Experimental results with ceramic components in passenger car diesel engines. In: *Ceramic materials and components for engines*, Bunk, W. (ed.), Bad Honnef, Verlag Deutsche Keramische Gesellschaft, 1986, pp. 1155–64.
5. Zernig, N., Brennraumisolation von Dieselmotoren durch den Einsatz keramischer Werkstoffe. In: *Einsatzschancen keramischer Werkstoffe im Motorenbau*, Conference paper, Haus der Technik e.V. (ed.), 1987.
6. Körkemeier, H., Erfahrungen mit Portlinern im PKW-Otto-Motor. In: *Einsatzchancen keramischer Werkstoffe im Motorenbau*, Conference paper, Haus der Technik e.V. (ed.), 1987.
7. Mielke, S. and Sander, W., Keramische Komponenten zur Verringerung des Kühlungsbedarfes von Kolben. In: *Einsatzchancen keramischer Werkstoffe im Motorenbau*, Conference paper, Haus der Technik e.V. (ed.), 1987.
8. McLean, A. F., Materials approach to engine component design. In: *Ceramic materials and components for engines*, Bunk, W. (ed.), Bad Honnef, Verlag Deutsche Keramische Gesellschaft, 1986, pp. 1023–34.
9. Fingerle, D., Gundel, W., Olapinski, H., Insulation of exhaust gas channels in combustion engines. In: *Ceramic materials and components for engines*, Bunk, W. (ed.), Bad Honnef, Verlag Deutsche Keramische Gesellschaft, 1986, pp. 1135–44.
10. Colombo, K. and Lanzavecchia, G., New ceramics: a strategic view—industrial, economic and societal impact. In: *Materials science monographs: high tech ceramics*, Vincenzini, P. (ed.), Amsterdam, Oxford, New York, Tokyo, Elsevier, 1987, pp. 3–30.
11. Rothman, E. P., Clark, J. P. and Bowen, H. K., Ceramic turbocharger cost modeling and demand analysis, *Int. J. High Technol. Ceram.*, **3**, 1987, 63–78.
12. Johnson, L. R., Tectia, A. P. S. and Hill, L. G., *Structural ceramics research program: a preliminary economic analysis*, Center for Transportation Research Report No. ANL/CNSN-38, Argonne National Laboratory, March 1983.

10

Ceramic Heat Engine Programs in the USA

RICHARD C. BRADT*

Department of Materials Science and Engineering, University of Washington, Seattle, Washington, USA

ABSTRACT

Ceramic heat engine programs in the USA are discussed in general with a philosophical approach, within the perspective of the framework of the DOE/ ORNL program. From the technical viewpoint, ceramic processing, design considerations and the establishment of a ceramic data base are all considered to have made substantial progress within the past decade. It is, however, re-emphasized that the economic consideration is an equally important one for the ultimate utilization of technical ceramics in consumer heat engines.

1. INTRODUCTION

Heat engine ceramic programs in the USA are now entering their second decade of intense research and development activity. During that time period, the understanding of design criteria and the improvement of ceramic material properties have experienced significant advances in every imaginable theoretical and practical sense. Within the last decade, the progress of technical ceramics in general has been unparalleled in the long history of that class of engineering materials. Yet the commercial ceramic heat engine, which many enthusiasts had predicted in the 1970s would be commonplace by now, continues as an

*Present address: Mackay School of Mines, University of Nevada-Reno, Reno, Nevada 89557, USA.

elusive spectre that always seems to remain just over the horizon. However, technical achievements have established for the first time a fundamental basis for the utilization of engineering ceramics into many demanding engine applications, a basis that was totally lacking at the onset of the era of ceramic engine programs nearly two decades ago.

The excitement which has been generated by the concept of a ceramic engine has a recent parallel in the discovery of superconducting ceramics, although the enthusiasm level for the latter appears to far exceed the ceramic 'fever' designation which was initially applied to the ceramic heat engine activity. Nevertheless, it is appropriate to consider these two different applications of technical ceramics within the perspective of the realities of commercialization for consumer products as P. E. Ross[1] has done in his *New York Times* article from which the next four paragraphs are excerpts.

Investors intrigued by the promise of the new superconducting ceramics might do well to consider the painfully slow progress that engineers are making in developing another class of ceramics, one they hope will sharply improve the performance of diesel engines.

A few years ago, promoters were predicting the imminent commercial use of ceramics in high-temperature diesel engines. The stability of ceramics — some expand much less than metals when heated — was one attraction. More important; though, some ceramics provided such good insulation that design engineers had visions of increasing diesel engine temperatures from today's level of under 400 °C to 1000 °C, a rise that would help burn fuel more efficiently.

In addition, the engineers said, ceramic engines would generate hotter exhaust gases, which could be used to produce more energy. The light weight of ceramic parts would also increase efficiency because less energy would be needed to move them. In theory, ceramic engines would also require less energy for cooling and lose less energy to friction between various parts.

Unfortunately, as the National Research Council concluded last month, such advanced diesels will not be practical in the near future because of various problems with materials and design. Some problems, including the brittleness of the ceramics that have been considered for the engines, are the same that bedevil those who want to manufacture superconducting ceramics.

Ross's article continues in a popular science tone, explaining the technology of ceramic materials. However, his initial paragraphs contain several vital messages that are worthy of further consideration, especially relative to the utilization of ceramics in heat engines. These messages are: (i) the rate of commercialization, (ii) unrealistic promises or expectations, and (iii) fundamental scientific barriers. In any new technology it is worth considering these three items. Each will be addressed within the perspective of technical ceramics in general and the use of engineering ceramics in heat engines in particular.

The rate of commercialization is a legitimate concern with all new technologies. It is appropriate to draw parallels with ceramic cutting tools, a prime example in the field of engineering ceramics, although perhaps not as demanding an application as many engine components and certainly not your average consumer product. An approximate date for the 'discovery' of ceramic cutting tools is the mid-1950s, but it is only within the past couple of years that ceramic cutting tools have truly achieved commercial prominence. The 30 intervening years almost seem excessive, yet it is a typical period for product commercialization. Not only are there necessary iterations in the evolution of the ceramic material itself, but there are often also necessary parallel technical improvements in the 'system' which utilizes the discovery.

In the example of ceramic cutting tools, it was the parallel evolution in the machine tools themselves (lathes, milling machines, etc.) that enabled the eventual widespread acceptance of ceramic tooling. Frequently the entire 'system' which utilizes the discovery must also be substantially improved or drastically redesigned and there is no reason to expect that it will be any different for the utilization of ceramics in heat engines. There are no reasons whatsoever to believe that one-for-one substitutions of ceramic engine components for the currently commercial 'metallic' ones will be possible in engines as they are now designed. Perhaps some radical new type of engine design will be necessary for commercial ceramic engines to finally appear in the family car? No matter what promises engineering ceramics have for the future, no one should be misled to believe that any of those promises will be fulfilled on a widespread basis before the turn of the century, the year 2000.

Within the perspective of the rate of commercialization, there are several other factors which merit strong consideration in the transition of any technology from the laboratory to consumer applications. One is the time required for educators to introduce the benefits of the technology to the practicing engineers who will actually prescribe the technology. Often a complete generation of students is required to pass through academia. Another is for the concepts to 'filter down' to some of the secondary and tertiary applications design personnel, many of whom have no basic understanding of the scientific phenomena which are involved. Newspapers are often helpful in this latter information transfer.

Unfortunately, newspapers are also the culprits which kindle the fires of unrealistic expectations. There have been more than a few of these fantasies in the era of ceramic heat engine fever. A couple of examples will suffice. The adiabatic engine concept is one, which in its extreme envisions virtually no energy losses to a cooling system and dramatically increased combustion chamber temperatures accompanied by overall engine efficiency improvements, perhaps approaching the 50% level. It is now almost universally recognized and also accepted that combustion chamber insulation will require extensive and elaborate technologies to extract the energy from the hotter exhaust gases and even then increased efficiencies will be much lower than many have fantasized. The low heat rejection engine (the real description of the adiabatic engine) may have advantages, but that of dramatically increased fuel efficiencies is certainly not one of them.

Another unrealistic expectation is that ceramic engines will not require any form of lubrication. Erroneous perceptions exist that the friction between ceramic/ceramic surfaces is so low that practically no wear occurs and lubrication is unnecessary. If ceramic engines contact surfaces do operate at temperatures near to 1000°C, as opposed to only a couple of 100°C, then entirely new lubricant systems will actually be required. It will necessitate major lubricant development programs, perhaps comparable in scope to the ceramic heat engine program itself. These two examples should certainly suffice to illustrate a few of the obvious problems with unrealistic expectations.

The final message in the Ross article introduction also merits serious consideration, although it is not completely clear at this point in time just what all of the fundamental scientific and technological barriers may be, as many have not yet been identified. However, as should be expected, the brittleness problem will remain a serious one for the fabrication and the application of consumer ceramic heat engines and heat engine components, just as it has always been with all ceramics.

Relative to the brittleness, it seems certain that monolithic ceramics will never achieve the fracture toughness levels that are commonplace with most metallic materials. Within current metal-oriented engine designs this is a very serious problem. However, true design modification instead of one-for-one ceramic for metal substitutions will surely alleviate some of this difficulty. Joining of ceramic to metallic components may also be expected to present formidable challenges as thermal expansion and elastic property mismatch are a basic fact of nature. Wear problems, particularly whenever moving ceramic and metallic parts are in direct contact, will provide for additional problems and certainly push lubricants to well beyond their current limits, particularly their upper level temperatures of performance.

These focal points of the Ross article introduction may seem unduly pessimistic, but should not be viewed as such. They are in fact brutally realistic and must be accepted within that perspective if ceramic components are to assume a significant role in the heat engines of the 21st Century.

2. US CERAMIC HEAT ENGINE PROGRAMS

Ceramic heat engine programs in the USA are a complex intertwined mixture of federally and industrially funded activities. Major efforts are in progress at NASA facilities, Wright-Patterson AFB and at the national laboratories such as Argonne and Oak Ridge. All of the major automotive manufacturers have their own substantial in-house programs and are simultaneously involved in cooperative efforts with ceramic manufacturers and universities. The innovative cooperative ceramic manufacturer/automotive parts supplier efforts such as the TRW/ Norton and GTE/Eaton joint endeavors are becoming increasingly focused, while turbine manufacturers such as Allison and Garrett have also been quite active. It is nearly impossible to identify a ceramic manufacturer that is not participating in some form of ceramic heat engine program. New companies such as Adiabatics, Inc. are even being formed to participate in the activities. Substantial efforts continue.

It simply is not feasible, nor even possible, to review all of those ceramic heat engine programs within the limitations of this meeting, much less within this single presentation, thus it is necessary to adopt a general philosophical approach to these activities, referring to items and materials from other sources. Two reviews have been recently published by the NRC/NAS/NAE/ICM group, one on the 'State of the art and projected technology for low heat rejection engines'[2] and the other on 'Ceramic technology for advanced heat engines'.[3] Both of these studies are quite realistic and perhaps may be termed moderately critical of some portions of the US activities. Of course, any reviews or studies of these types may be expected to contain some criticisms. The

TABLE 1
DOE/ORNL university R & D projects (k $)

Title	FY-83	FY-84	FY-85	FY-86	FY-87
Powder Char MIT (Univ.)	30	30	30	0	0
Powder Scale Up (Univ.)	0	0	0	50	300
TT Al_2O_3 Mich (Univ.)	0	70	60	160	300
Low Exp Ceramic VPI (Univ.)	0	0	0	150	150
Coat Adher Test UT (Univ.)	0	0	18	0	0
Life Prediction (Univ.)	0	0	0	0	340
Static Fatigue U ILL (Univ.)	0	0	120	120	120
Env Effect U Dayton (Univ.)	102	125	125	163	150
Fract Tough U Wash (Univ.)	61	60	150	115	120
Hi Temp Tensile (NCA&T Univ.)	0	300	100	200	200
NDE/Char (Univ.)	0	0	0	0	0
Totals	193	585	703	958	1 540

latter of these two reviews is a particularly interesting one for it primarily addresses the coordinated long term effort which has been sponsored by the Department of Energy (DOE) and is monitored by ORNL, the Oak Ridge National Laboratory high temperature ceramics group.

The DOE/ORNL 'model' program has been in progress for the past 5 years. It appears to be the best coordinated of all of the US programs, past and present. The funding of specific efforts for that time period is summarized in Tables 1–3 as university, government and industrial laboratory projects. In addition, modest sums were also allocated to program management and to technology transfer, the latter focusing on an annual meeting in Detroit. The goal of the program has been advertised to assist with the development of an adequate industrial ceramic technology base to provide reliable and cost effective ceramics for heat engines. The allocation of nearly 70% of the project funding to industrial participants seems very appropriate. Unfortunately, the industrial financial statements are not available as to the parallel or matching fund type research efforts that those companies have also allocated to the specific and related research endeavours. As nearly all of the industrial contractors are involved in ceramic heat engine programs of their own, it is clear that a substantial leveraging of the listed funding exists. The industrial aspects of the DOE/ORNL program appear to be balanced, varied and excellent.

The remainder of this presentation will address the overall US

TABLE 2
DOE/ORNL government laboratory R & D projects (k $)

Title	FY-83	FY-84	FY-85	FY-86	FY-87
DOE Suppt Contracts (Govt)	128	10	0	100	0
SIC Powder ORNL (Govt)	60	60	0	0	0
SIC Whiskers LANL (Govt)	0	60	0	0	0
Sintering SI_3N_4 AMTL (Govt)	0	70	70	70	70
Disp Tough SIC ORNL (Govt)	70	70	77	0	0
Oxide Mat Comp ORNL (Govt)	0	200	345	355	355
Sol Gel Oxide Pwd ORNL (Govt)	90	105	100	100	100
Inj Mold Comp ORNL (Govt)	0	100	200	277	300
CVD Coating ORNL (Govt)	154	200	0	0	0
M-C Brazing ORNL (Govt)	0	90	250	280	280
Design Allow Code AMTL (Govt)	0	130	0	0	0
Adv Statistics ORNL (Govt)	0	25	40	60	50
Failure Analysis NBS (Govt)	0	0	100	0	0
Phy Properties ORNL (Govt)	0	0	0	120	60
Char TT Ceram AMTL (Govt)	0	70	80	100	100
Time-Temp Prop AMTL (Govt)	0	100	85	0	0
Frat Behavior ORNL (Govt)	130	230	190	166	200
Cyl Fatigue ORNL (Govt)	0	0	90	280	280
Life Pred Meth ORNL (Govt)	0	0	0	50	0
Std Tensile Test NBS (Govt)	0	0	90	90	100
NDE Dev. ORNL (Govt)	0	0	70	150	150
NDE Assessment ORNL (Govt)	0	0	80	50	0
Matl Char AMTL (Govt)	0	0	130	130	130
C Tomography ANL (Govt)	0	50	0	0	0
Totals	504	1 660	1 997	2 278	2 175

programs within the perspective of the goals of the DOE/ORNL program, namely the development of reliable and cost effective ceramics for components of heat engines. That choice has been made for this presentation as the DOE/ORNL endeavor is an overall consumer oriented program that is not focussed on the military or space. The specific DOE/ORNL program base is such that three broad areas are funded within the categories: (i) materials processing, (ii) design methodology and (iii) data base and lifetime predictions. The approximate allocation to each of those three areas, all of which are absolutely vital to the success of the extensive utilization of ceramics in heat engines has been nearly 60%, 10% and 30% for the DOE/ORNL program. This division of resources may not be the same for each of the other individual efforts within the USA, but it certainly provides a

TABLE 3
DOE/ORNL industry R & D projects (k $)

Title	FY-83	FY-84	FY-85	FY-86	FY-87
SIC Powder SOHIO (Ind.)	150	100	110	106	0
SIC Whiskers (Ind.)	0	0	0	150	225
Sintering Furnace GEO (Ind.)	0	80	80	70	70
SI3N4 Powder FORD (Ind.)	150	100	110	139	0
Powder Scale UP (Ind.)	0	0	0	50	1 660
Processing Monolithics (Ind.)	340	156	0	0	1 000
Whisk Tough SI3N4 SC AIR (Ind.)	130	0	35	337	0
Trans Tough SI3N4 RYDNE (Ind.)	110	130	40	220	0
Whisk Tough SI3N4 IM GTE (Ind.)	0	165	0	335	0
Composite Dev (Ind.)	0	0	0	100	250
Adv Composites (Ind.)	0	0	0	0	2 000
Syn Proc TT Ceram Norton (Ind.)	120	75	100	366	0
Layered Comp Ceramatic (Ind.)	0	0	150	85	0
SIC Whisk-Mullite GEVFSC (Ind.)	120	0	130	180	0
Adv Coating AGT (Ind.)	0	0	100	150	150
CR203 Coat Eval Cummins (Ind.)	120	100	0	0	0
Wear Resist Coating (Ind.)	0	0	200	150	175
Coating Scale UP (Ind.)	0	0	0	0	0
M-C Joint AGT (Ind.)	0	0	100	200	200
M-C Joint Diesel (Ind.)	0	0	100	100	150
C-C Joint AGT (Ind.)	0	0	100	200	150
Hi T Coat GTEC (Ind.)	77	111	146	11	0
Dyn Interface BCL (Ind.)	122	125	110	189	150
Wear Behavior (Ind.)	0	0	0	0	0
Adv Statistics GE Lab (Ind)	0	200	101	116	100
Translucense Effects ITI (Ind.)	0	50	0	0	0
Exp Life Test FORD (Ind.)	0	230	0	0	0
Life Prediction (Ind.)	0	0	0	75	600
Char Dev (Ind.)	0	0	0	200	700
Totals	1 439	1 622	1 717	3 529	7 580

perspective as to where the government/industrial/university consensus believes that the funding should be allocated. The three areas will be addressed in a philosophical sense in addition to the summary one.

From the materials processing viewpoint there exists a widely held opinion that ceramic components for heat engines cannot be produced uniformly, nor perhaps even economically. Some engineers believe that the properties of monolithic ceramics will never be adequate and that ceramic composites will be necessary. Related technical areas such as joining and surface modification are also pertinent to ceramic utilization. Any serious analysis of the use of ceramics in heat engines will almost certainly conclude that technical advances are necessary in the ceramic materials processing area. Certainly the failure of the advanced gas turbine (AGT) program to succeed using domestic ceramic materials heralded a real need for increased ceramic processing research in the USA. Current heat engine ceramic programs certainly do emphasize ceramic processing, perhaps too much in the government laboratories and universities, for it is in the industrial research and development laboratories as well as those industrial production facilities that the 'new' processing must ultimately be applied. Nevertheless, the basic concepts which are being studied in the government and university research laboratories are probably necessary for the required processing advances. However, success will only be realized when the ceramic producers such as Coors, GTE, Norton and Sohio, for example incorporate the 'new' processing into their manufacturing to improve the reliability of the ceramic products and to substantially reduce their costs. Although many engineers are highly skeptical of the use of ceramics in heat engines because of the brittleness and reliability issues, the cost consideration is every bit as serious a challenge, if not more so! In some instances the processing costs must be reduced by more than an order of magnitude for ceramics to be competitive with metals.

As for ceramic composites, there is no longer any question that vastly increased toughnesses are possible. However, the question for their application in commercial heat engines is an economic one. There exists legitimate cost concerns for the utilization of monolithic ceramics in heat engines, so that it must be concluded for the present that only military or space applications will be able to afford ceramic matrix composites. Only a revolutionary manufacturing process, as opposed to the normal evolutionary ones, could make ceramic composites economically attractive at this time, or in the near future.

Design covers a wide range of items from the ceramic material microstructural design to the actual structural design of the components by mechanical engineers. It is the latter which needs the greatest attention and perhaps the most radical innovation, if ceramics are to truly be successful in consumer heat engines. It is also the technical area that appears to be receiving the least support in all of the US heat engine programs. It is probably the most critical. A technically sound design methodology has evolved that incorporates the elements of fracture statistics, finite element analysis and probabilistic approaches. It has, unfortunately, often been applied to one-for-one substitution schemes of ceramics for metals. More often than not, catastrophic failures have resulted. However, substantial progress has also occurred. As with all designs these failures and the resulting iterations are a vital component of the design process. Unfortunately, ceramics are in the position that their design iterations of only a decade must compete with those of metals for nearly a century. It is a formidable challenge!

While there will undoubtedly be substantial progress in the engine design area, probably only by the ceramic heat engine manufacturers themselves, any truly innovative design advances may be long in achieving acceptance. That wait may be an extended one, as a new generation of design engineers that have not been influenced by the accepted concepts and pedagogical prejudices may be needed. It is a difficult task to escape from these influences, but it may be a prerequisite, for ceramics do not appear to be capable of simply replacing metals within heat engine designs that have evolved over such an extended period to optimize the properties of metals. Many engineers believe that a radical new design concept for a heat engine is the boost which ceramics need to truly succeed in the consumer arena.

The third component of the DOE/ORNL program is that of the development of a long term data base that engine designers can rely upon for property values. Modern technical ceramics are not a handbook property variety of materials, so a reliable data base must be established. Practically none exists! Its almost a 'Catch-22' situation as neither the ceramic material producers nor the engine manufacturers wish to absorb the costs, yet designers, who are perhaps the most conservative of all engineers, must have long term property data so that failure can be avoided. Establishment of a suitable data base not only encompasses the pedestrian gathering of routine numbers, but also requires related activities such as the development of new NDE/NDT techniques and the coupling with the ceramic processing and design procedures. It is nearly a Herculean task, but an absolutely necessary one if ceramics are to be truly successful in commercial heat engines. While technical ceramics appear to suffer from the almost total absence of a data base, perhaps it is a blessing that the opportunity exists to start from the beginning and develop that data base in a coherent manner, one that properly interfaces with the other requirements. It is refreshing that the DOE/ORNL program recognizes the importance of a lifetime data base and it is encouraging that both producers of ceramic engines and the engine manufacturers themselves seem to be supporting its establishment.

3. SUMMARY

Ceramic heat engine programs continue with enthusiasm in the USA, as not all ceramic researchers have turned to superconductivity. The DOE/ORNL ceramic engine program is an excellent model example of the necessary industry/government/university cooperation in our quest for a ceramic heat engine. While there is not a commercial ceramic engine available for consumers to order quite yet, many of the technical and economic problems have been positively identified and substantial progress has been made toward their solutions.

It is well known that prototype ceramic engines have been designed and built and that today's technical ceramics can function in some 'modified one-for-one' replacements for metals, but a commercial ceramic heat engine is not yet commonplace. It is probably premature to expect it, as we have only recently transcended the initial stages of the usual time period for technological development of the concept in a true commercial sense. The next decade should be an exciting one in that respect, for we have now only passed the threshold of the iterations necessary for commercial success of the ceramic heat engine.

At the present there have already been many successes for technical ceramics in heat engine components and on today's automobiles in general. Perhaps there are as many as 50 ceramic parts now in production for the average passenger automobile, whereas only a couple of decades ago applications were limited to the glass headlamps and windshields, and the alumina sparkplugs. Although many of these applications are for non-structural ceramic components with sensor, optical and electrical functions, 'structural' ceramic applications such as valves, rocker arm wear pads and bearings are rapidly increasing. With each of these, both the ceramic and automotive communities are learning more about ceramic processing and engineering design with ceramics and they are accumulating the lifetime data base, which is so desperately needed. However, the present commercial ceramic applications on automobiles all have in common the virtue of cost effectiveness. Naturally, their feasibility and their reliability were proven long in advance, but they did not appear on the family car until the cost effectiveness was firmly established. That is certainly a necessary and critical requirement for the commercialization of the ceramic heat engine.

REFERENCES

1. Ross, P. E. The long struggle to harness ceramics, *New York Times*, 16 September 1987.
2. *State of the art and projected technology of low heat rejection engines*, National Academy Press, Washington, D.C., USA, 1987.
3. *Ceramic technology for advanced heat engines*, National Academy Press, Washington, D.C., USA, 1987.

11

Developments in the UK Engineering Ceramics Programmes

D. A. PARKER

T & N Technology Ltd, Rugby, UK

AND

G. W. MEETHAM

Rolls-Royce Ltd, Derby, UK

ABSTRACT

The complementary DTI-supported Advanced Ceramics for Turbines (ACT) and Ceramic Applications in Reciprocating Engines (CARE) programmes share a common origin and maintain close liaison in areas of common interest.

Description of the fundamental limitations of metal turbine blades leads on to the identification of silicon nitride and carbide as potential successors. The key areas of technology necessary to develop ceramic blades are identified as stressing/design methodology, increasing temperature capability and defect tolerance, more fundamental lifing philosophies and improved manufacturing processes. Developments in these areas under ACT and demonstration of ceramic turbine blades, turbine shroud rings and air bearings in a Rolls-Royce gas turbine are reported.

Work under the CARE programme is divided into four sections, namely materials substitution, thermal insulation, turbochargers and materials development. Potential improvements are described for reducing wear rates and in high temperature lubrication; also for reducing particulate emissions and cooling system volume by combustion chamber insulation, for decreasing the inertia of turbocharger rotors and valve gear by using the lighter ceramic materials and for reducing manufacturing costs by development of improved processing.

1. INTRODUCTION

Both of the current collaborative programmes involving the UK government and industry in the development of engineering ceramics for engine applications began with an initiative by Rolls-Royce PLC in February 1983. Initially both internal and external combustion engines were included in the terms of reference, but later two separate programmes emerged in recognition of the inherent differences between the systems. Thus problems of the gas turbine, with its external combustion, continuous burning and high operating temperatures were addressed by the Consortium for Advanced Ceramics for Turbines (ACT), whereas those of the internal combustion engine, with its intermittant burning, reciprocating motion and generally lower operating temperatures, were addressed by the Consortium for Ceramic Applications in Reciprocating Engines (CARE). It has, however, always been recognised that there is a large area of common interest between these programmes. In consequence careful arrangements have been made to exchange information and co-ordinate projects in this common area. Projects undertaken by both of the Consortia are supported by the Department of Trade and Industry at the rate of up to 50%.

2. CERAMICS IN GAS TURBINE ENGINES

The major objectives in the design of advanced gas turbine engines will continue to be reduced fuel consumption, reduced costs of ownership and, in the case of aero gas turbines, increased power/weight ratio. When translated into materials technology, the requirements are essentially increased strength, reduced density and increased temperature capability.

Nickel superalloys have been the prime high temperature turbine materials from the outset, and their progressive development has been most impressive. There is however, a fundamental limitation on the further development of nickel superalloys — their melting temperature of some 1300 °C. The need therefore is for materials with an inherently higher temperature capability, both from the need to operate at higher turbine temperature and because of the improvement in efficiency which would result from the use of materials which do not require internal air cooling (Fig. 1).

Many materials, largely non-metallic, meet this requirement but

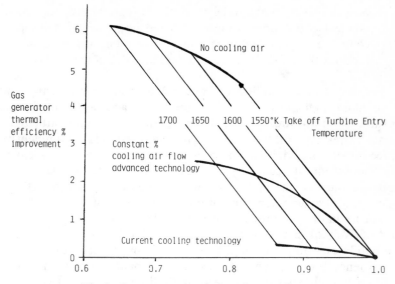

Fig. 1. Gas generator relative air mass flow.

most are eliminated from serious consideration because of excessive oxidation or inappropriate physical and mechanical properties. The physical properties of inductile materials are of prime significance in gas turbine operation with its rapid thermal fluctuations. Of the ceramic materials currently available, silicon nitride and silicon carbide have the potential for the earliest application in gas turbine engines. These materials are stronger than nickel superalloys above about 1000°C although some forms have inadequate strength at intermediate temperatures to be considered for components such as turbine blades. They have superior creep strength and oxidation resistance, are potentially cheaper than nickel superalloys and contain no strategically significant elements.

For acceptance in the gas turbine engine, ceramic components must be capable of demonstrating a reliability in operation at least as good as that of the metal components they replace. This requires the further development of four key areas of technology in which knowledge is currently incomplete, but the necessary advances can be defined and progress is being achieved. These areas are:

— stressing/design methodology;
— materials with increased temperature capability and defect tolerance;

— material behavioural understanding to progressively supplement and replace the current empirically based life prediction philosophy;
— component manufacturing processes which will consistently produce the required microstructure and defect level.

Because of the inability of ceramics to deform plastically and redistribute high local stress, it is critical to accurately predict the magnitude of the stress within the component, including particularly contact stresses at interfaces. This prediction requires comprehensive three-dimensional thermal and stress analysis including finite element techniques that will accurately analyse the whole component. Associated software must include advanced mesh generation programmes and appropriate capacity for input of boundary conditions.

The traditional empirical approach to life prediction is based on the use of Weibull statistics to predict stress levels for high surviving probabilities, with the Weibull Modulus m being a measure of strength consistency. It has been widely used with reasonable success in general terms at low temperatures. There are limitations, however, to its general applicability. There is presently little evidence to indicate the accuracy of predictions based on extrapolation well beyond experimental data, and the theories make no allowance for interaction of defects. Clearly interactions can occur. Agglomerations of small defects can coalesce and act as a single larger defect. At high temperatures additional effects can occur involving microstructural or chemical changes to the material. These include localised plastic flow inducing crack healing or generating creep voids, oxidation which can produce crack healing or introduce new defects, and microstructural changes such as grain growth, phase changes or surface evaporation. Thus several potential high temperature effects can influence strength, and the original defect population present at ambient temperature cannot necessarily be regarded as controlling the mechanical response of a material to stress at high temperature. This emphasises the limitations inherent in the empirical life prediction approach, and ongoing work to increase the behavioural understanding of ceramic materials is essential for the development of a satisfactory life prediction methodology.

Material composition, structure, properties and behaviour are ultimately linked with the component manufacturing process. The use of silicate liquids in the densification process in the manufacture of most silicon nitride-based ceramics is well established. Because regions of uncrystallised glass residues after sintering can provide sites for creep cavity nucleation (and thus increase creep strain) there is a compromise between sinterability and properties. Temperature capability can be augmented by reducing the remaining liquid volume to below the critical size for cavity nucleation, or by selecting glass compositions capable of complete crystallisation on post-sintering heat treatment. Reaction sintering is an established process for silicon nitride and carbide and results in a very small dimensional change, the disadvantage being lower strength levels compared with materials obtained by the alternative routes. A combination of both types of process may, however, be attractive. The requirement for increased defect tolerance is being addressed through the development of ceramic composites, with reinforcement materials such as whiskers. The incorporation of whiskers and fibres in silicon nitride, carbide and other materials adds another dimension of difficulty to the development of manufacturing processes with the capability of producing net-shape components with the required microstructure and defect level.

Rig and bench engine testing of ceramic components had been carried out in the UK during the decade prior to the initiation of the ACT programme some 5 years ago. Recent demonstration has included ceramic turbine blades, turbine shroud rings and air bearings in a Rolls-Royce bench development small gas turbine engine.

The ACT Consortium involves UK engine manufacturers, component suppliers, research establishments and Government Departments in a co-ordinated series of programmes which are jointly funded by the Department of Trade and Industry and the companies involved. Organisations carrying out these programmes include T & N Technology, AE Turbine Components, Tenmat, Lucas Cookson Syalon, AERE Harwell, Fairey Tecramics, BCRA and Pilkington. The programmes are concentrating on material and process developments aimed at the needs of the user industry. The necessary associated developments in the understanding of material behaviour and in component stressing and life prediction are being addressed in the user companies.

In the monolithic ceramic area, the basic chemical and physical understanding involved in turbine blade manufacture by injection moulding and CIP/green machining is well advanced, as is the interaction of material composition and HIP processing. Component manufacturing techniques have been developed for Syalon 201. Grinding and lapping techniques for close tolerance component regions are at an advanced development stage. Earlier UK work on Nicalon reinforced pyrex has been extended and with $0.49V_f$ fibre content, a flexural strength of 1·25 GPa, Weibull Modulus of 30 and work of fracture values in excess of 50 kJ m^{-2} are available. The indications are that these effects can be developed in higher temperature ceramic systems.

3. CERAMICS IN RECIPROCATING INTERNAL COMBUSTION ENGINES

The CARE Consortium has 28 industrial members collaborating on 13 running projects with a total committed spend of just under £5 m. The Consortium is divided into four Sections, three dealing with potential areas of application of ceramics to reciprocating engines and the fourth with materials development. The Substitution Section deals with the developing and testing of ceramic components with a view to replacing some conventional materials currently used in engine manufacture. The Engine Insulation Section is developing ceramic components to reduce heat losses to coolant and the Turbocharger Section the UK capability to manufacture turbocharger components from ceramics. The Consortium is thus user-led, and as programmes in these three areas develop, the requirement for further materials development can be increasingly refined. However, the specialist knowledge of the members of the Materials Section also helps to define both desirable material properties and the means of achieving them.

3.1 Materials Substitution

One of the longer term studies being undertaken by the Consortium is the basic tribology of ceramic materials rubbing against themselves or against metals. The use of ceramics to reduce the heat losses from diesel engines, or to reduce their cooling requirements,[1,2] poses major tribological problems. Cylinder wall temperatures of 350–550°C[3] may result from the use of well-insulated cylinder liners, leading to a near absence of liquid lubrication due to evaporation or carbonisation. Thus with such high degrees of insulation, alternative approaches are being considered including gas lubrication and the use of solid lubricants. However, even at normal engine operating temperatures well established tribological data are needed, since many of the applications proposed for ceramics rely upon the low wear rate attributable in part to their hardness. As part of the Materials Substitution activity these data are being systematically collected as a function of material, surface finish, lubricant and temperature.

In another Materials Substitution project, a ceramic pre-combustion

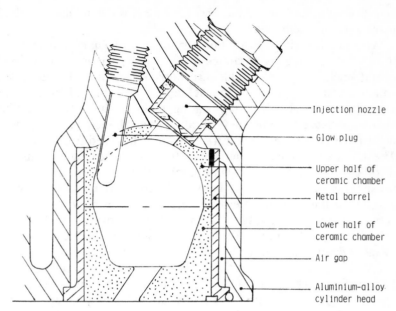

Fig. 2. Cross-section of complete ceramic swirl chamber.

chamber is being developed as an alternative to the traditional nickel based superalloy chambers in current use. A particularly attractive goal is the reduction of particulate emissions due to the more complete combustion expected from the higher operating temperatures. Indirect injection diesel engines equipped with ceramic chambers[4] have met the 1987 model year US Federal Emissions Standard of 0·2 g/mile particulates for passenger cars.

This particular design of pre-combustion chamber is illustrated in Fig. 2. Extensive material selection tests revealed that silicon nitride was the best material for manufacture. The chamber is made from two ceramic parts held together by a shrink-fitted metal barrel. This assembly technique serves both to equalise the temperature distribution over the surface of the ceramic (thereby reducing the thermal stresses therein) and to place compressive stresses in the ceramic to offset any tendency to move into tension. It is interesting to note that additional insulation is provided by the air gap surrounding the metal barrel. Another function of the barrel is to provide a means of taking up differences in thermal expansion between the ceramic and the metal cylinder head. Stressing of the whole assembly was carefully modelled using finite element techniques.

A further Materials Substitution project, led by Jaguar Cars, concerns the use of engineering ceramics in the valve gear of a four-valves/cylinder petrol engine. Introducing ceramic valves and tappet inserts in combination with polymeric bucket tappets and spring retainers will significantly reduce the reciprocating mass of the valve actuation system. Increases in efficiency, durability and refinement, along with decreases in engine noise and friction, are anticipated from this work. The alternative tappet materials to be investigated include silicon nitride, silicon carbide and sialon, each with three alternative surface finishes (as-fired, ground and lapped). Silicon nitride and sialon were chosen for the manufacture of the ceramic valves by virtue of their superior toughness. Testing will commence in a fatigue rig and a single cylinder engine; later durability evaluations will be conducted in a six-cylinder engine first on the test bed and then in a vehicle.

Another project concerned with valve gear is being co-ordinated by Perkins Engines and seeks to avoid lubrication other than at the cam/tappet interfaces, through the use of ceramic bearing materials. This project will thus benefit particularly from the results of the fundamental studies. Wear tests are under way to evaluate a number of materials including sintered silicon nitride, sintered silicon carbide, a toughened alumina and two zirconia ceramics. For more lightly loaded components, such as valve guides, reaction bonded silicon nitride will also be evaluated. The latter material has the great advantage of virtually no change in volume during nitriding, so that near-final machining may be undertaken in the green state. Designs have been produced for several valve train components employing ceramic inserts, mostly incorporated by shrink fitting.

Lastly in the Substitution Section, a project on in-cylinder components seeks to improve valve/valve seat compatibility in medium speed diesel engines by the use of ceramics for one or more of these components.

3.2 Thermal Insulation

As mentioned in the previous section, much of the early work on the application of ceramics to reciprocating engines concerned cylinder insulation and diversion of heat from the coolant to an energy recovering device in the exhaust pipe as a means of increasing overall fuel efficiency. Whilst this concept remains as an ultimate goal, more immediate gains can be made by capitalising upon the smaller cooling system required to cope with the reduced rejection of heat to coolant. Such a project is being conducted within the CARE Insulation activity and led by Leyland-DAF. Two engines have been built with ceramic coated pistons, valves and cylinder heads and incorporating minimally cooled block and head structures.[2] The 33% reduction in heat-to-coolant allowed both the radiator and cooling fan to be reduced in size. The overall consequence was that whereas no gain in fuel economy was noted on the test bed, an improvement of 4% was realised in vehicular operation. Although the pistons tested to date have been insulated by a ceramic coating, other pistons incorporating monolithic ceramic and air-gap insulated pistons with ceramic coating are also being prepared for test.

To obtain maximum benefits from an insulated combustion chamber the passages leading to the turbocharger, and indeed to any power generating device situated in the exhaust pipe, would also need to be insulated. Another project, led by Fairey Tecramics, aims to develop the product and process aspects of exhaust port liners. A number of materials containing aluminium titanate are being evaluated for this application, including reaction sintered and pre-formed versions of the material. In parallel with work on property improvement, a pilot production route for the aluminium titanate is being developed. Exhaust port liner shapes based on a current truck diesel engine have been produced and are to be incorporated shortly in casting trials. The thermal durability of the several candidate materials is being assessed using simplified concentric tubes.

The two other projects in the Insulation Section are of a more fundamental nature. One seeks to quantify the benefits and disadvantages of thermal insulation with regard to such engine performance parameters as fuel efficiency, emissions, fuel tolerance and cost. It will thus provide useful background in interpreting the results of the other three insulation projects, and be of particular benefit in guiding future work. The fourth project will use a combination of test rigs and engines to provide a fundamental understanding of the high temperature gradients and thermal stresses set up by engine insulation. In particular it will investigate and seek to optimise the use of glass ceramics and plasma sprayed coatings to insulate engine components.

3.3 Turbochargers

Work in the Turbocharger Section is concentrated on two projects, one concerned with vehicular applications and the other with much larger engines. Manufacturers of petrol engines have for some time been using turbocharging as a method of power boosting that does not require a radical redesign of the engine. It is thus particularly suitable for enhancing or extending a model range. Currently most turbocharger rotors are manufactured from high chromium steels and superalloys such as Inconel 738. A major limitation in the current design of petrol engines is that of a relatively slow speed of response due to the high inertia of the rotor. Substitution of ceramic for the above metals can reduce the rotor weight by up to two thirds with a corresponding improvement in response time. Furthermore ceramic rotors maintain their strength at significantly higher operating temperatures than metal rotors. Greater fuel economy has been claimed in petrol engines and reduced exhaust emissions in diesel engines.[5]

Ceramic materials under consideration include silicon nitride and silicon carbide. Several Japanese manufacturers, e.g. NTK for Nissan and Kirocera for Mitsubishi now produce silicon nitride rotors for commercial use. Manufacturing techniques include slip casting and injection moulding, whereas hot isostatic pressing has been invoked as a means of increasing component reproducibility and reliability. Tests at Garrett of turbochargers with silicon nitride rotors have shown durabilities in excess of 150 000 km and blade tip speeds have been in excess of 500 m s^{-1}. Manufacture of these articles in, albeit modest, commercial quantities will have allowed a considerable quantification of the true cost of maintaining quality in a component whose failure mechanism is probabilistic. Work on the CARE Small Turbocharger Project will allow the corresponding information to be derived from UK sources. However, the project is structured so that the results will also be relevant to other small gas turbine applications.

Applications of ceramics within the Large Turbocharger Project include novel heat shields and arrangements for insulating the casing.

3.4 Materials Development

The objectives in the Materials Section of the CARE Consortium are two-fold: namely, to assist in providing the most appropriate ceramic materials for current applications and to develop new materials for future use with superior properties or at reduced cost. In response to the second objective a project is being led by Unitec Ceramics.[6] Electro-refined zirconia powders have been developed with sinter activities comparable to those of chemically co-precipitated materials. These electro-refined PSZ powders can be used to produce dense engineering

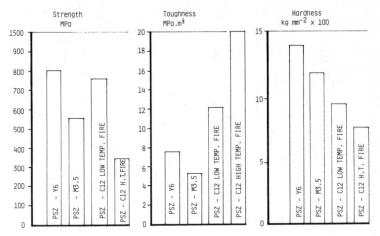

Fig. 3. Typical mechanical properties.

ceramics with good mechanical properties. The Ceria-PSZ system in particular produces extremely tough monolithic ceramics with high Weibull Moduli. By controling the chemistry of this system, particle size and firing schedules, the ceramic can be engineered to give optimised properties in a manner analogous to metal alloy property control. Typical mechanical properties are illustrated in Fig. 3.

Although it is too early to be definitive, the simpler approach to refining employed in this project augers well for cost reduction relative to material produced by chemical routes. As will be discussed later on, the cost of ceramic components is likely to be a critical factor in determining their use in reciprocating engines. The project is now moving on to develop agglomerated powders for coatings which will then be applied and evaluated; also to develop fabrication and test methods for monolithic components.

In a second Materials Project, on novel thermal barrier coatings, an attempt is being made to design the crystal structure of non-zirconia coatings to achieve the required properties.

4. DISCUSSION AND CONCLUSIONS

Possible timescales for the introduction of engine ceramic components[7] are set out in Table 1. Thus although some ceramic components are already commercially available, much product and process development will be required before all the components listed can be

TABLE 1
Timescales for the introduction of engine ceramic components

Time scale (years)	Diesel/Petrol	Gas turbine
0–5	Precombustion chambers Tappet shims and similar surfaces in valve gear Turbochargers Coatings (piston rings, valves and seats) Port liners	Glow plugs Precombustion cups Turbocharger rotors and housings
5–15	Lightweight valve gear Coatings (erosion, corrosion) Monolithic ceramics Insulation (piston crowns, exhaust ports)	Component for prototype engines — Shrouds, combustors Regenerators
15+ engine redesign	Minimum cooled engine, Minimum friction bearings, valves, valve seats, piston/liner systems	All ceramics AGT

TABLE 2
Current UK prices of some ceramic components

Component	Ceramic (£)	Metal (£)
Turbocharger rotor	20–25	3
	100 (accounting for rejects)	
	Trigger price 6	
Valve-guides	3–5 times cost of metal part	
Push rod		<0·5
Tappet		<0·5
Cylinder liner	170	
	60 (plasma spray coating)	
	13 (chromium oxide coating)	
Valve seat	6 (small volume)	0·4

incorporated. Since the relationship between capital and operating cost is different in the gas turbine and in small reciprocating engines, so the relationship between enhancement of performance and acceptable increase in price is correspondingly different.

In the aircraft gas turbine operating costs are paramount. Thus avoidance of the need to supply cooling air for metal blades, for example by using ceramic blades, provides a powerful commercial incentive for expenditure on ceramic development. In the small reciprocating engine on the other hand, initial purchase price is critical. Table 2[8] gives some idea of the current disparities between ceramic and the corresponding metal components. It must be borne in mind that it is the total cost of the processes required to bring the product to the customer that must be considered in judging the feasibility of the application. These costs would be a reflection not only of the inherent cost of the materials themselves, but also of the quality control and other procedures needed to ensure consistently satisfactory operation. Nevertheless, notwithstanding these constraints, the application of new high temperature materials to gas turbines and reciprocating engines provides an exciting yet fundamental challenge to the innovative materials engineer.

REFERENCES

1. Kamo, R. and Bryzik, W., Adiabatic turbocompound engine programme SAE Paper 810070, Society of Automative Engineers, Warrendale, Pennsylvania, 1981.

2. Wonnacott, E. J., *The application and value of low heat loss engines to military vehicle cooling systems*, Seminar on Low Heat Loss Engines, Institution of Mechanical Engineers, London, July, 1986.
3. Toyama, K., Yoshimitsu, T., Nishiyama, T., Shaimauchi, T. and Nakagaki, T., Heat insulated turbocompound engine, SAE Paper 831345, Society of Automotive Engineers, Warendale, Pennsylvania, 1983.
4. Ogawa, Y., Ogasawara, T. and Machida, M., Complete ceramic swirl chamber for passenger car diesel engine SAE Paper 870650, Society of Automotive Engineers, Warrendale, Pennsylvania, 1987.
5. Lasker, M. K. and Byrne, J. L., *Experience with ceramic rotor turbochargers.* Conference on Turbocharging and Turbochargers, Paper 4, Proc. No. MEP-246, Institution of Mechanical Engineers, London, 1986.
6. Blackburn, S., Hepworth, M. A., Kerridge, C. R. and Senhenn, P. G., *Toughened zirconia ceramics from electro-refined powders*, United Ceramics Ltd., Stafford, 1987.
7. Sandberg Howe, C., New ceramic material gives champion performance, *Ceramic Industry,* **127**(5A), 1986, 34–6.
8. Kirk, J. N., Ceramic components in automotive applications, *Metals and Materials,* **13**(11) (Nov. 1987), pp. 647–52.

12

The Present Attitude towards High Technology Ceramics in Japan

HIROSHIGE SUZUKI

Emeritus Professor, Tokyo Institute of Technology, Tokyo, Japan

ABSTRACT

Current Japanese research and development work in the field of high technology ceramics is duscussed, and the important trends are described.

Attention is given to the areas of materials processing and evaluation, and to some new materials and applications. Prospects for the near future for technologies and products are reviewed.

1. INTRODUCTION

The author has written many reports on research and development programmes and the status of progress on various new ceramics (in Japan, they are generally called 'Fine Ceramics').[1,2] In these reports, the author also referred to some of the existing problems and the future prospects. However, the progress in technology systems in leading industries is remarkable, and the change of environment enclosing high performance ceramics (hereafter referred to as 'High Technology Ceramics') is also great. Therefore in this report the way of thinking on, and of carrying out, research and development of high technology ceramics in Japan will be briefly explained again, and recent changes in the systems and organizations concerned with research and development of high technology ceramics will be described. In addition, results of research and development in the field of ceramic processing and application technology will be introduced and problems regarded as important will be listed; the means of solving them, and some results of the research into solutions for the problems will be examined and forecasts for the future on solutions to these problems will be made.

2. THE WAY OF THINKING ON, AND OF CARRYING OUT, RESEARCH AND DEVELOPMENT OF HIGH TECHNOLOGY CERAMICS IN JAPAN

As one of the groups of supporting materials for the high technology industries (although the amounts used are not necessarily large) ceramics are required to have high functional properties in a range of application areas, and are thought to have become so important that they can not be substituted by other materials.

In order that materials can be used as parts or components, they must have a desired shape, a desired dimension, and the required microstructure, so that the necessary properties and performance can be attained. To manufacture such products, the development of processing technology is required. In addition, this technology should be supported by other approaches such as 'evaluation technology' (the acquirement of a data base, and testing techniques), and 'application technology'.

In Japan a national project concerning research in and development of Fine Ceramics, aspects of which had previously been carried out independently, was drawn up under an initiative of the Ministry of International Trade and Industry of Japan (MITI) in 1981, 10–20 years behind those in the UK and the USA. The project, called 'JISEDAI (next generation) project' for short, is an ~10 year project aimed at the establishment of the basic technology concerning the manufacture of high strength, highly corrosion-resistant and highly wear-resistant materials, mainly comprising SiC and Si_3N_4. The author has reported on this project in other publications, and accordingly this paper will point out only later developments in this area.

It is noteworthy that in the 'Jisedai' project, concrete aims such as the manufacture of all-ceramic gas turbines or heat engines, as in the USA or the FRG, are not defined.

Besides the national project, research and development in high technology ceramics has been being carried out freely by many enterprises, national and public institutions, and universities, by small groups, and by large numbers of smaller groups involving cooperative

TABLE 1
Expectation for hi-tech ceramics

(1)	A material of new dimensions for high-tech industry
(2)	Stimulation of existing industry
(3)	Creation of a new major industry
(4)	Savings in energy and scarce metal resources
(5)	Contribution to growing information intensive industry in regional communities

work by makers and users on a considerable scale. Among these some have recently succeeded in achieving satisfactory results and in putting them to practical use.

Needless to say, besides research and development in Si_3N_4 and SiC ceramics, research and development in sialons, AlN, ZrO_2, mullite, ZrB_2, BN, B_4C and other materials has also been carried out. Furthermore, studies on coatings and composites of these materials have been conducted steadily.

As described above, various high-technology ceramics have already achieved a position as one of the groups of important new materials. Additionally, some interesting social effects are expected. These are shown in Table 1.

3. AN OUTLINE OF HIGH TECHNOLOGY CERAMICS RESEARCH AND DEVELOPMENT ACTIVITIES IN JAPAN

There is a considerable number of ceramics researchers and engineers in Japan. Research and development institutes and organizations to which they belong are listed in Table 2. Of these, ERAHPC is an organization carrying out the 'JISEDAI' project, in which researchers and engineers belonging to fifteen industrial firms are collaborating closely with each other. In addition six governmental research institutes and laboratories of four universities are giving assistance in the study of fundamental problems in new ceramics.

With regards to the initial results of research and development in this project, the reader is referred to other reports.[1] Recent achievements will be reported in the following section.

Japan Fine Ceramics Center (JFCC) was established in 1985 at Nagoya with the assistance and cooperation of industrial, public and

TABLE 2
Organizations for research and development of new ceramics in Japan

The Ceramic Society of Japan (CSJ)
Japan Society for the Promotion of Science (JSPS), No. 124 & No. 136
    subcommittee
Japan Fine Ceramics Association (JFCA)
Japan Fine Ceramics Centre (JFCC), in Nagoya City
Engineering Research Association for High Performance Ceramics (ERAHPC)
National Institute for Research in Inorganic Materials (NIRIM) in Tsukuba
Government Industrial Research Institute, Nagoya (GIRI, Nagoya)
Government Industrial Research Institute, Osaka (GIRI, Osaka)
Government Industrial Research Institute, Kyushu (GIRI, Kyushu)
More than 15 Universities
More than 50 private enterprises

academic sectors. The object of JFCC is to carry out research and development on the manufacture and application of fine ceramics, and to contribute to the improvement in quality of fine ceramics and their expanded use and application. As part of this work, the establishment of standardized test methods and the development of evaluation systems are involved. A great deal of effort has been made to attain these objects. In addition, based on these objectives, JFCC has been carrying out contract research and testing, and holds meetings for the international exchange of information. In April 1987, new buildings and research facilities were completed and more than fifty research staff belonging to nine research groups are now carrying out studies on a range of research and development projects.

Japan Fine Ceramics Association (JFCA), is well known, and was established by ninety-six members (mainly consisting of industrial enterprises) in 1982 with the intention of promoting the development of the fine ceramics industry through information exchange between firms, universities and nations. It has since grown steadily, and the number of its member companies now exceeds two hundred. This is a result of the recognition by many companies of the future development of fine ceramics, and of the far-reaching effects of their research and development, as described in the previous section (Table 1). Researchers belonging to many research institutes and universities are participating in the research and development of fine ceramics according to their positions and preference. Needless to say, many basic and theoretical studies have been carried out in universities. Recently, interest in diamond, cubic boron nitride, new glasses and new carbon materials, has been heightened, and forums for these materials have been founded.

4. PRESENT STATUS OF THE DEVELOPMENT OF MANUFACTURING TECHNOLOGY

The field of high-technology ceramics contains not only structural ceramics, but also electronic ceramics and bioceramics. In the manufacturing technologies some methods and techniques are specific to the individual ceramics, although common techniques also exist. For instance, in the electronic ceramics area, since there are many kinds of special electroceramics such as thin films and microcomponents having a special grain-boundary structure, and those requiring mass production techniques, methods and techniques for the manufacture of materials have been developed corresponding to these particular requirements. However, in this section, the author will deal mainly with structural ceramics.

4.1 Results of Studies on the 'JISEDAI' Project

The results of studies in the first phase of the 'JISEDAI' project have been reported.[1] In short, methods for the synthesis of raw materials have been established with respect to silicon nitride and silicon carbide ceramics; forming and sintering of test pieces ($10 \times 10 \times 10$ mm) have been successfully carried out; target properties for these ceramics have been achieved; the technical knowledge relating to these ceramics has been stored. In addition, fundamental studies on the hipping and gas pressure sintering processes have been extensively developed in this phase of the project. The second phase of the project will be completed by the end of 1987. The results are being summarized and will be presented in 1988. According to an interim report, the results are as shown in Table 3.

The third phase begins this year (1988). As one of the targets of this project the development of methods for increasing fracture toughness of fine ceramics will be added in the third phase.

In the following a brief explanation of Table 3 is given:

(i) Si_3N_4 and SiC synthesized raw powders were delivered to ERAHPC members, inspected for properties, used for the manufacture of various kinds of simple experimental com-

TABLE 3
A few remarkable achievements in the second phase of 'JISEDAI' project

(a) Material powders could be synthesized on pilot scale mass production
(b) High anti-oxidation property was attained in Si_3N_4 (0·25 mg/cm^2 after 1000 h at 1300°C)
(c) High joining strength between Si_3N_4 and steel ($\sigma_{4P} = 510$ MPa at 400°C)
(d) Improvements in various process handlings and determination of optimum conditions, clarification of governing principles
(e) Proof testing and analytical method under difference in test and practical stress modes
(f) Observation of crack propagation under high temperature SEM

ponents, and evaluated by the members. As a result, many kinds of raw powders were found to be satisfactory. At the same time, an economical consideration of each synthesis process was made.

(ii) A high strength material, a corrosion resistant material, and a high precision, wear-resistant, material were made on an experimental basis by the use of various forming/sintering

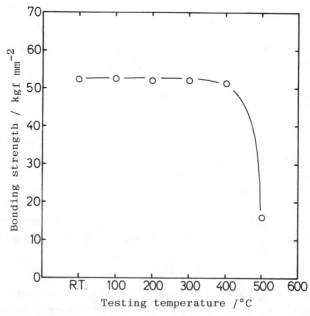

Fig. 1. Bonding strength of Si_3N_4/S-45C at high temperature.[3]

techniques such as injection moulding/hipping (IM/HIP), naturally sintered/hot isostatic pressing (ns/HIP), and reaction sintering/gas pressure sintering (RS/GPS). In addition, methods of mixing powders with sintering aids, and methods of pretreatment, were also investigated. The properties of fine ceramics made by these techniques, and the optimum conditions for their manufacture were clarified; (b) is an example of these.

(iii) Ti/Cu/Ag was coated on the surface of a Si_3N_4 ceramic in the form of a multilayer coating by a high frequency ion plating method. The coated Si_3N_4 was heated at 1000°C for 5 min to firmly bond the Ti/Cu/Ag layers to the surface of the Si_3N_4. Steel (S-45C) was then soldered to the Ti/Cu/Ag plated Si_3N_4 by inserting between them a superalloy having a low thermal expansion coefficient. The relation of the bonding strength of the joined specimen to temperature is shown in Fig. 1.

(iv) Since (e) and (f) have no direct relation with the manufacturing techniques of ceramics, the explanation is omitted.

4.2 An Example of the Results of Research and Development on High Technology Ceramics (Excluding those of the 'JISEDAI' Project) in Japan

Research and Development on high technology ceramics have been carried out in many firms and laboratories. The present status will be described as follows.

(i) Raw material powders were at first imported from foreign countries for experimental use. However, nowadays some companies in Japan have come to produce, and sell, these raw material powders with qualities equal to or higher than those from foreign countries. Not only members of ERAHPC but also many companies belonging to other technical branches such as chemical and textile, are participating in the research and development of ceramics. Some of them have already succeeded in manufacturing high quality raw material powders. For instance, one firm is producing Si_3N_4 powder on the large scale of 100 t per year, and it sells various brands of Si_3N_4 classified by its crystal form (such as α or β), particle size, specific surface area, particle shape (such as spherical shape or whisker) and purity. Almost the same may be said with regard to fine particle SiC.

In addition, the production of ZrO_2 powder is increasing. One company is producing 200 t per year of ZrO_2 powder. Further, several companies are producing pure ZrO_2, or ZrO_2 stabilized or semistabilized by addition of Y_2O_3, CaO, MgO, CeO_2 and so on. Raw materials for Al_2O_3–Y_2O_3–ZrO_2 ceramics containing up to 20% Al_2O_3 are also being produced. In the preparation not only of ZrO_2 ceramics, but also most ceramics, the selection and the homogeneous mixing of the sintering aids is known to be very important. As a result, raw material powders in which the selection and mixing of the sintering aids have been carried out by manufacturers during the raw material synthesis stage and processed so that the users have only to purchase, form and sinter them, are commercially available on the market.

Besides these, various kinds of raw material powders such as those of AlN, BN, sialon, B_4C, TiB_2, mullite, Al_2O_3, TiO_2 and AlON, as well as Al_2O_3, have been produced. Furthermore, production on an industrial scale of whiskers and fibres of ceramics such as SiC, Si_3N_4, Si(Ti)CN, SiO_2–Al_2O_3 and carbon has also become possible, in order to allow these materials to be put on the commercial market in the 1990s.

(ii) Forming, sintering and coating processing studies have also been carried out by national laboratories and industrial firms, and good results have been obtained. As a forming method, injection moulding is excellent, but the merits of slip casting have been rediscovered lately, and this method has been used successfully for forming large sized components. As one of the great successes among these, a method developed by NIRIM, and used for the manufacture on a commercial basis by the NGK Spark Plug Co. Ltd, with the financial help of the Research Development Corporation of Japan (JRDC) is referred to. In this method Si_3N_4 is formed by injection moulding or cold isostatic pressing (CIP) and then subjected to two-stage gas pressure sintering. By the use of this method, turbocharger ceramic rotors having Weibull modulus $m = 16$ have been successfully made.

As for sintering, densification mechanisms during the initial stages are understood to a considerable degree, but final stage mechanisms have not yet been clarified. According to our recent experiences the final stage is very important. This is because the object of sintering is not only to make the compacted green body as dense as possible, but to improve such properties as K_{Ic} or thermal conductivity by controlling the microstructures and the grain-boundary structures of the ceramics. In other words, if the grain growth is excessively inhibited in order to increase strength, K_{Ic} will become too low. Accordingly, K_{Ic} can be made considerably larger, and can be increased to the order of 6 MPa m$^{1/2}$ by making the grains grow to some extent, and, if possible, by causing needle-like or plate-like crystals to develop. Such examples are found in Si_3N_4, sialon, Al_2O_3 and SiC. Considerable amounts of glass phase usually remain in the grain boundaries. The glass phase, as is well known, lowers the strength of ceramics at high temperature, and at the same time, in most cases, it also lowers the thermal conductivity. Attempts to avoid a reduction in strength at high temperature by crystallization of the glass phase in the grain boundaries of Si_3N_4 have been made over many years. One noteworthy investigation on sintering recently carried out in Japan is a study on high thermal conductivity AlN. Although already reported,[4] the essential points are shown in Fig. 2. Namely, when Y_2O_3 and

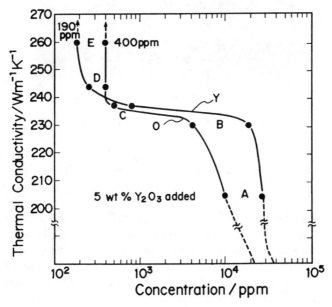

Fig. 2. Impurities (O_2 and Y) concentration versus thermal conductivity of AlN sintered at 1900°C.[4]

(a)   (b)

Fig. 3. Cutting edges of (a) diamond coated and (b) uncoated insert.[5] (Cutting conditions: speed, 500 m/min; feed, 0·1 mm/rev; depth of cut 1 mm; time, 5 min.)

Fig. 4. Surface of aluminium alloy after cutting test.[5]

YF$_3$ is added to AlN and sintered, initially the green compact is densified by a liquid phase-sintering mechanism. In the final stage, impurities contained in the raw materials come together in the grain boundaries, and are discharged through the grain boundaries. Some impurities seem to form a solid solution in the AlN crystals. As a result, very fine SEM images in which little grain boundary phase exists have been obtained. Incidentally these sintered materials have a very high thermal conductivity of approximately 260 W m^{-1} K^{-1} at room temperature, and have begun to be put to practical use as a substrate for integrated circuits.

It is important to consider chemical vapour deposition as a kind of sintering. Especially a method of forming a ceramic matrix in the gap between fibres by chemical vapour infiltration (CVI) in manufacturing ceramic matrix composites should be noted. Further, CVI is promising as a method for coating with ceramics, and for synthesizing ceramic powders.

As one example of a recent brilliant success in Japan, let me now introduce the formation of diamond film from methane and hydrogen. The fundamental study was carried out in NIRIM and the technique was used for industrial production by Mitsubishi Metal Co. Ltd, under the auspices of JRDC. Figure 3 shows the cutting edges of diamond coated and uncoated inserts prepared by coating diamond on a super-alloy. Figure 4 shows the appearance of the surface of an Al alloy cut with the cutting tool.

Diamond film prepared in Seiko Electronic Co., Ltd, in a similar way showed a high thermal conductivity of 900–1000 W m^{-1} K^{-1} as expected. Although research and development of diamond for utilization as a semiconductor at high temperature has not yet been completed, diamond has a great future as a raw material for high temperature semiconductors, as well as where high hardness and high heat conductivity are required.

(iii) One of the biggest problems in manufacturing ceramics is the difficulty in machining materials such as Si_3N_4, SiC and ZrO_2. The machining cost often amounts to 10–100 times the raw material cost. Therefore, it is essential that a method for the preparation of sintering materials having a near-net-shape is adopted so that the machining cost is minimized. However, in order to make a component the first requisite is that cutting and grinding of materials can be carried out easily. For this reason, in Japan the development of machines having high rigidity and a multi-mechanism capability (involving electric discharge, electrolysis, vibration and laser) has been planned. Further, some companies are making fine ceramics electrically conductive, and carrying out electric discharge machining. Among these, a method of processing with a machining centre equipped with a cast iron bonded diamond grinder invented by Professor T. Nakagawa[6] is worthy of attention.

Recently, various machinable ceramics which can be used at high temperature have been developed.[7] Figure 5 shows an

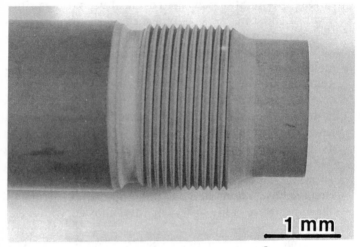

Fig. 5. A machinable SiC ceramic.[7]

example of machinable SiC based ceramic materials tested by Niihara.[7]

5. EVALUATION TECHNOLOGY FOR ENGINEERING CERAMICS

Since in engineering ceramics brittle fracture occurs, and strength varies considerably, the mistrust of users for these materials is increasing. This is the main reason for the lack of expansion in potential markets for engineering ceramics. As a countermeasure it is necessary to measure the desired properties and behaviours of ceramic products as accurately as possible, accumulate these data, and construct a data base for users to utilize in designing products. In order to achieve this objective, it is desirable first to establish and standardize methods of measurement, and next to develop methods for inspecting non-destructively, and evaluating flaws on the surface or in the interior of components, and to systemize these methods.

In addition, in order to make up for the deficiencies of the methods, proof testing is necessary. At present, many companies are carrying out proof tests in the form of total inspection. These techniques of examination and evaluation are called 'evaluation technology'. In this article only the present status on the methods of non-destructive inspection of ceramics will be described.

5.1 Non-destructive Testing (NDT) of Fine Ceramics

The strength of fine ceramics tends to be low because of their low fracture toughness. The size of the flaws to be detected is very small, and is thought to be one-tenth to one-hundredth of those in metals, that is, of the order of $10–100\,\mu m$. Accordingly, the methods conveniently used for metals cannot be used unchanged for fine ceramics. The history of the study in this field in the world is short, and the level of study is not yet sufficient. But since the manufacturing technology of ceramics has reached a considerable level, and fine ceramics have been put to practical use such as in motor car parts, the importance of non-destructive testing (NDT) and non-destructive inspection (NDI) is rapidly increasing. In the 'JISEDAI' project, a fundamental investigation and study has been carried out in its second phase, and photoacoustic spectroscopy (PAS) is being given attention. It is reported that flaws having a size of about $50\,\mu m$ existing on the surface of Si_3N_4 and SiC ceramics have been detected. However, with this method there are some problems such as that of the length of time taken for the detection of flaws.

For the detection of internal flaws, the X-ray computed tomography (CT) technique has been investigated. But since this method is insufficiently insensitive, a study is being carried out on enhancing the sensitivity of the detector to make a high precision internal flaw-evaluation instrument, with an X-ray generator having a small focus point.

Studies on NDT are being carried out in many industries and research institutes. Toyota Central Research and Development Laboratory has developed a computerized ultrasonic system for NDT of fine ceramics and put the system to practical use in the manufacturing line of ceramic precombustion chambers. The detection sensitivity is reported to be about $50\,\mu m$ for surface flaws, and $160\,\mu m$ for internal flaws.[8] Nippon Steel Corp. has developed a system for the detection of surface flaws of about $40\,\mu m$ diameter, and internal flaws of about $80\,\mu m$ diameter, automatically by a surface wave method and a C-SAM system using transducers of 25 MHz point focus type.[9] JFCC has carried out an investigation on NDT systems for fine ceramics since 1986, at the request of the Mechanical Social System Foundation, Japan. A summary of the methods adopted for study and the results obtained is

TABLE 4

Comparison of detectability in NDI methods studied for some kinds of seeded flaws, with remarks

Methods	Detectability μm (%)				Test time	Remarks	
	Surface or sub-surface flaws		Internal flaws			Advantages	Disadvantages
	Notch	Knoop-mark or pore	Void	W-wire			
Microfocus X-ray film (CRT)	10 (0·3) 20 (0·4)		80	30–50 (2·3–4·3)	Long Short	Intricate shape OK No couplant automatic	• thick specimen → low resolution • very thin crack • some inclusion
Ultrasonic high frequency AC Scope	50–10	200	80–100		Short	Surface: OK Internal: OK Automatic	• sub-surface → dead zone • couplant
SAM	10	20	<30	<30	Long	High resolution	Simple shape Flat surface
SLAM	10	150	30	110	Long	High resolution	
PAS	20				Long	No liquid couplant	
Fluorescent dye penet.	10	100			Middle	Simple equip. Cheap	Qualitative

shown in Table 4. In this study, Si_3N_4 ceramics were used as the standard sample, and the flaws were seeded artificially. Machined notches, and marks and cracks by Knoop indentation were used as surface flaws. Internal flaws were made by inserting tungsten and polyamide resin balls and filaments in the ceramic green body before sintering. Various methods such as those using X-rays, ultrasonic waves, scanning acoustic microscopy (SAM), scanning laser acoustic microscopy (SLAM), fluorescent dye penetration, and acoustic emission analysis, have been studied. The effects of artificial flaws on the bend strengths of samples of Si_3N_4 ceramics have been measured. The results are shown in Fig. 6. The results on flaws in SiC and Al_2O_3 obtained by other researchers are also shown in Fig. 6.

As shown in Fig. 6 and Table 4, it is almost possible to detect flaws, of a size thought undesirable for engineering ceramics, by NDT methods

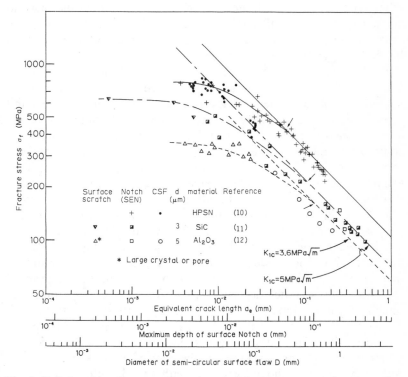

Fig. 6. Relationships between flaw size and fracture stress of various fine ceramics, Si_3N_4, SiC and alumina.[10–12]

which have, to a considerable degree, been brought to completion. But each method is not perfect. Further improvement in detectability and resolving power is needed. Although SAM and SLAM are excellent in these respects, problems such as cost of the equipment, scope of application, measuring time, and systemization still remain. Research and development in this field will be carried out more actively in future in Japan.

6. PRESENT STATUS OF HIGH TECHNOLOGY CERAMICS PUT TO PRACTICAL USE, AND THEIR PROSPECTS

The high technology ceramics which have been most put to practical use in Japan are the electronic ceramics. The market volume of electroceramics has reached several hundred GYen per year. On the other hand, the market volume of ceramics for mechanical, thermal and chemical use is still small and is less than one fifth to one tenth of that of electronic ceramics. The market volume of ceramics for the optical and biotechnology industries is also still small.

But because of the unique characteristics of ceramics as described above and the improvement in quality caused by remarkable progress in their manufacturing techniques, a tendency for the market to expand continuously is seen. Many practical application examples, having accomplished several tens of times the 'merits' (=life × quality/cost) of conventional materials as the result of the ingenious use of ceramics, are seen in mechanical seals, and in ceramic lined (or coated) impellers of exhausters for high temperature exhaust gases containing dust, etc. In this section let us introduce briefly the present conditions of some examples of high technology ceramics used for mechanical and thermal purposes. Examples from the auto-industry are shown in Table 5.

High technology ceramics have been used successfully for cutting tools, dies and mechanical seals for a long time. What should be mentioned specially is that parts exposed to severe vibration, high speed and heat impact, such as swirl chambers for diesel engines and impellers of turbochargers, could be made successfully, and the market is becoming stable (up to the present, about three million glow plugs, and thirty thousand turbocharger rotors are in the market, with the rotors recently being produced at the rate of ~10 000 per month). Although not yet on the market, a car with an adiabatic piston engine was developed and is being tested by Isuzu Motors Co. Ltd (this car has

TABLE 5

Examples of high-technology ceramic materials used in the automotive industry

Name of component	Materials
Glow plug	Si_3N_4
Pre-combustion chamber	Si_3N_4
Turbocharger turbine rotor	Si_3N_4
(Rocker arm insert	Si_3N_4)
(Exhaust vale	Si_3N_4)
(Cylinder head	ZrO_2 or Si_3N_4)
Honeycomb catalysts carrier	$MgO\text{-}Al_2O_3\text{-}SiO_2$
Mechanical seals of water pump	Al_2O_3
Insulator	$Al_2O_3 \cdot SiO_2$ fibre
Piston head ring	$Al_2O_3 \cdot SiO_2$-Al alloy
Sensors, heaters and electronic ceramics	

already run 5000 km). Further, a car with an adiabatic diesel engine is being tested by Komatsu Mnfg Co. Ltd for durability. Since ceramic parts of small size and simple shape are easy to make, and reliability dependence on properties such as component strength is thought to be high, ceramics were initially put to practical use in such small parts. Additionally a firm with many years experience has succeeded in the manufacture of a product of large size (one metre or more) and attracted public attention. Some Si_3N_4-SiC ceramics of very high purity and of large size are in use in the semiconductor industry. Ceramics of the same quality may also be used in high temperature heat exchangers.

Figure 7 shows high alumina ceramics in use in the field of precision machines and measuring instruments. High alumina having the same quality as this is used for surface plates, V-block and similar applications. These tools have been given a favourable reception because of their excellent wear resistance, high dimensional precision (of the order of sub-μm), corrosion resistance, stainless properties, high rigidity, and durability. A turbo-molecular pump having a casing of an Al alloy, and a shaft and blades of Si_3N_4 was made on an experimental basis. Since the pump was non-magnetic and of excellent corrosion resistance, it was used in JT-60 (an experimental plant for nuclear fusion) and the result was satisfactory.[13]

Fig. 7. Precision two-dimensional measuring machine with ceramic guideway (Al_2O_3 ceramic) (courtesy of TOTO Ltd).

Ceramics are in competition with Inconel alloy or Ti-aluminide (intermetallic compound) for machine parts when these parts are used in the range of 600–900 °C. But, when they are used at temperatures over 900 °C or in environments causing oxidation or corrosion, ceramics are thought to be at an advantage. In other words, for the application field requiring materials having high hardness, high abrasion resistance, low density and low thermal expansion coefficient, it is thought that ceramics will rapidly come to be used in the future. The market volume of high technology ceramics at present (partly involving estimates) is shown in Table 6.

7. UNSOLVED TECHNICAL PROBLEMS, AND ASSESSMENTS OF THESE PROBLEMS

Many problems with ceramics are still unsolved. Objectives to be realized during the next stage are summarized in Table 7. Very many results of experiments carried out to solve these problems have been reported. Since the situation has not much changed, a brief supplementary explanation will be given as simply as possible to avoid duplication.

TABLE 6
Market volume of high-tech ceramics and its forecast to 2000

Kinds of ceramics	1985(A)	(1987)	Estimated in this year		2000(B)	(B)/(A)
			1990	1995		
Electric/electronic (cont. magnetic)	600·8	(651·3)	869·5	1 249·9	1 787·5	3·0
Mechanical	106·0	(112·1)	135·5	183·9	252·0	2·4
Heat resistant	58·0	(57·7)	110·0	204·8	422·3	7·3
Bio-industrial	59·5	(59·2)	70·2	94·0	176·9	3·0
Optical	32·2	(43·5)	134·8	406·0	1 007·1	31·3
Remainder (nuclear, super cond.)	1·2	(1·3)	4·9	163·1	491·0	410
Total	857·7 (A)	(925·1)	1 324·9	2 301·7	4 136·8 (B)	4·8

Japan Fine Ceramics Association.
Unit: BYen; March 1987.

TABLE 7

Objects of the research and development on fine ceramics at next stage (for auto-gas turbine engine)

1.	Increase reliability	
	Weibull modulus m from 16 to 20 K_{IC} [MPa m$^{1/2}$] from 6 to 10~	
2.	For evaluations and applications	
	Standardization of test methods Accomplishment of NDE system Cost reduction	
3.	For much higher levels	
	Useable temperature	> 1 350–1 400°C
	Bending strength	> 750 MPa
	Creep	< 0.2% (1 000 h)
	Thermal shock resistance	> 300 K/s
	Corrosion and abrasion resistance more increased, etc.	

7.1 Increase in Weibull Modulus m, and Fracture Toughness K_{Ic}

In the first phase of the 'JISEDAI' project, as described above, ceramics having an $m > 20$ have been produced, and those having $m > 30$ were often obtained in the case of small size test samples. But in the case of components of more complicated form, or larger dimensions, or of products actually on the market, the value of m seems to be lower at ~16–17 or ~8–10. As to the cause, some factors are considered. One of these will be low fracture toughness K_{Ic}.

Since the importance of K_{Ic} has been recognized, many efforts will be increasingly concentrated on these problems from various industrial fields. As a means for enhancement of m the author proposed in previous papers the transformation of tetragonal ZrO_2 to monoclinic ZrO_2, reinforcement by fine particle dispersion, the dispersion effects of whiskers or fibres, utilization of in-situ reinforcement procedures (IRP) etc. Subsequently, many reports of studies of IRP, and of the confirmation of the effects of IRP have been made. Since such an interesting result was obtained for an experiment in fibre reinforcement, it will be introduced in the following. Figures 8 and 9 show flexure–load curves and bending stress–flexure curves of Al_2O_3 reinforced with waveshaped tungsten fibre obtained by M. Saito et al.[14]

From these figures, it is seen that the degree of damage and fatigue property is improved to a considerable degree by using shaped fibres

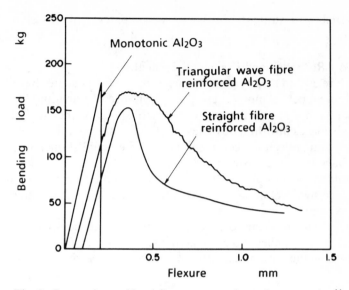

Fig. 8. Comparison of load-flexure curves in various samples.[14]

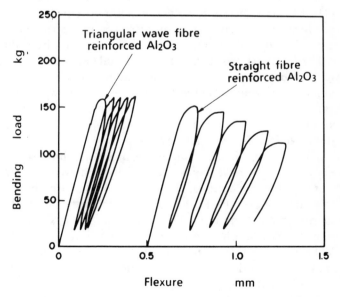

Fig. 9. Load–flexure curves for cyclic loading test.[14]

compared with straight fibres, and developments of this study are expected.

7.2 Cost Reduction and Evaluation Technology

With regard to these problems, the appropriate measures have been stated previously. Among these the importance of providing facilities to users through collaboration in establishing methods of design through the establishment of test methods, their standardization in making and collecting accurate data, and the establishment of a data base, have been increasing markedly.

Similarly, routes for reducing the cost, and improving quality assurance by developing and systemizing methods of NDI, have also been explored.

Since such a task cannot be carried out by private companies alone, JFCA, JFCC and other public research institutes are wrestling with the task in earnest in the form of industry–government–university collaboration.

8. CONCLUSIONS

It is important to realize that the ultimate goal for the Japanese National project (JISEDAI project) is not to manufacture specific final products, but to produce and evaluate high technology ceramics so as to establish the basic technology of ceramics for leading industries in to the next decade. The JISEDAI project has greatly stimulated many researchers and engineers in industrial firms, in national research institutes and in universities. Fine Ceramic technologies have thus shown remarkable development in a very small number of years. At present, many engineering ceramics, such as some automobile engine parts, have come into practical use, without any trouble.

The first objectives in developing the process technology of high technology ceramics have been successfully achieved, so the next objectives are clearly defined. These are to increase the fracture toughness values of $7 \sim 10$ MPa m$^{1/2}$, and to establish the standardization of measuring methods and of NDT systems, and others.

As for new ideas, there have been many interesting results of sintering and CVD experiments. Increasing fracture toughness is really a great problem. Research and developments carried out in Japan concerning the problem have been explained.

The scale of the present market of high technology ceramics has been stated and a prediction of its growth has been made.

ACKNOWLEDGEMENTS

The author gratefully acknowledges many friends who kindly gave their valuable data and were helpful in the preparation of this paper. The author also would like to express sincere thanks to many officers in MITI, and directors in JFCC who kindly agreed to the presentation of this paper. The author is also grateful to Dr F. L. Riley for advice on the final wording of the manuscript.

REFERENCES

1. Suzuki, H. Recent trends in the development of fine ceramics in Japan, *Mater. Sci. Engng,* **71** (1985) 211–26.
2. Suzuki, H. A perspective on new ceramics and ceramic composites, *Phil. Trans. R. Soc. Lond.,* **A322** (1987) 465–78.
3. Nakata, H., Honda, M., Miake, T. and Motoyoshi, K., Joining of ceramics to metal by multi-layer metallizing method, Pre-print No. 3-I-09, 1073–4 (1987) in Nagoya, Annual Meeting of Ceramics Society of Japan.
4. Isuge, A. and Shinozaki, K. Conditions of high thermal conductivity aluminium nitride, Inspec. 1986 winter No. 10 46–49 and private communication.
5. Kikuchi, N. and Komatu, T., *Characteristics of thin film growth in CVD diamond synthesis and application of the thin film synthesis technology for cutting tools,* ICHM-3, 3rd Int. Conf. on the Science of Hard Materials, held in Nassau, on 9–13 Nov. 1987, *Science of Hard Materials 3* Sarin, V. K. (ed.), Elsevier Applied Science, London, 1988.
6. Nakagawa, T., Suzuki, K. and Uematsu, T., High speed grinding of ceramics by machining centre, Ceramics Japan *(Bull. Ceram. Soc. Japan)* **21**(8) (1986) 704–10.
7. Niihara, K. *et al., Nanostructure and Mechanical Properties of SiC Consolidated Using Organosilicon Precursors,* Proc. of 3rd Int. Conf. on Ultrastructure Processing of Ceramics, Glasses and Composition, San Diego, Feb. 1987, in press.
8. Hayashi, K., Hatta, M. and Kawashima, K. Nondestructive inspection for void flaw in fine ceramics, pre-print No. 24 of 007, Special Research Committee, in the Japanese Society for Non-destructive Inspection, June 1987 in Tokyo.
9. Asai, K., Takeuchi, A., Ueda, N. and Kawamoto, J. Computerized ultrasonic system for ceramic pre-combustion chambers of automotive diesel engines, SAE Technical Paper Series 850534 Detroit, 1985.
10. Mechanical Social System Foundation Japan, Research report of inquiry on nondestructive inspection system for fine ceramics, 61-R-3, June 1987, Executed by Japan Fine Ceramics Center.
11. Takahashi, I., Usami, S., Nakakado, K., Miyata, Y. and Shida, S. Effects of flaw size and notch radius on strength of structural ceramics, *Yogyo-Kyokai-Shi,* **93**(4) (1984) 186–94.

12. Katayama, Y. and Matsuo, Y., Relationships between strength and flaw size of alumina ceramics, Pre-print of 3rd Discussion Meeting on High Temperature Materials, Jan. 1984, Kyoto, Japan, 34–38.
13. Mrakami, Y. *et al.*, *J. Vac. Sci. Technol. A.*, July–Aug. (1987) in press; and private communication.
14. Saito, M. and Mizoguchi, T., Improvement of fracture behaviour of brittle matrix composite, private communication.

Index